ENERGY AND CHARGE TRANSFER IN ORGANIC SEMICONDUCTORS

ENERGY AND CHARGE TRANSFER IN ORGANIC SEMICONDUCTORS

Edited by
Kohzoh Masuda
Osaka University
Toyonaka, Osaka
Japan

and
Marvin Silver
University of North Carolina
Chapel Hill, North Carolina

PLENUM PRESS • NEW YORK AND LONDON

Library of Congress Cataloging in Publication Data

U.S. — Japan Seminar on Energy and Charge Transfer in
  Organic Semiconductors, Osaka, 1973.
  Energy and charge transfer in organic semiconductors.

  1. Organic semiconductors—Congresses. 2. Energy transfer—Congresses.
3. Charge transfer—Congresses. I. Masuda, Kohzoh, 1931-   ed. II. Silver,
Marvin, ed. III. Title.
QC611.8.07U54   1973   537.6'22   74-7112
ISBN 0-306-30803-7

Proceedings of the U.S.—Japan Seminar on Energy and Charge Transfer
in Organic Semiconductors held at Osaka, Japan, August 6-9, 1973

© 1974 Plenum Press, New York
A Division of Plenum Publishing Corporation
227 West 17th Street, New York, N.Y. 10011

United Kingdom edition published by Plenum Press, London
A Division of Plenum Publishing Company, Ltd.
4a Lower John Street, London, W1R 3PD, England

Printed in the United States of America

Preface

Great progress has been made in the field of ordinary semiconductor physics and associated technologies. For the time being, if we could use new materials such as organic semiconductors progress in electronics could be accelerated.

Characteristics of organic semiconductors that are superior to others are: i) high photo-conductivity under irradiation along with low leakage current in the dark, ii) high sensitivity of the conductivity to various gases and to pressure. iii) possibility of using them in the amorphous state, iv) possibility of making devices of extremely small size, v) large variety of the materials, which makes suitable choice of material component easy.

A possible future development is a highly conductive material which could be used for electric power transmission – and which might help solve some of the problems posed by transmission losses.

The U.S.-Japan Seminar on Energy and Charge Transfer in Organic Semiconductors was held in Osaka Japan, 6-9 August, 1973. Completed results were summarized and the direction for the future was discussed. Information was exchanged quite freely and actively in a pleasant atmosphere. Many of the papers presented at the seminar are published here but unfortunately a few could not be included.

It would give us great pleasure if this seminar could be one step in the further development of the research in this field.

We wish to give our sincere thanks to the Japan Society for Promotion of Science and the National Science Foundation for financial support. Thanks are also due to Professor H. Akamatsu who gave us an opening address and stimulated comments throughout the seminar and to all participants.

Particular thanks go to Mrs. Y. Kido, Mr. T. Ugumori, Mr. T. Nishimura, Mr. K. Hiroyama, Miss E. Sakanaka, Miss K. Fujimoto, and Miss T. Tabata for their assistance throughout the seminar.

The editors have made minor changes without permission of the authors because of the importance of the rapid publication of the proceedings. We hope that we have not altered the original meaning. We accept full responsibility for any misconceptions introduced during editing.

Kohzoh Masuda
Marvin Silver
(Editors)

Contents

I. INTRODUCTION

OPENING ADDRESS

H. Akamatu

Department of Chemistry, Yokohama National University

Ohka, Yokohama

I am more than happy to meet together with you and to discuss our
common interests, the enrgy transport and the charge transport in
organic molecular crystals. We are grateful to NSF and JSPS for
the sponsorship of this meeting. I must express my thanks to Dr.
Silver the coordinator of U.S. side and to Dr. Masuda the organiz-
er of Japanese side, for taking trouble of promoting this meeting.To
such a meeting as this, of a group of specialists, I have nothing
to say as an opening address. However, if it is needed I would
like to do it in very few words.

It is more than twenty years ago when the electrical conduction in organic solids attracted our interests. In 1960 we met
together in Durham, North Carolina, and discussed on anthracene
crystal just as we do in this meeting. At that time, organic crystals meant a new field of investigation and we were excited facing
to this field.

Since then, many things have been known and many informations
have been accumulated concerning not only homo-molecular crystals,
but also inter-molecular compounds. After TCNQ had been found by
the people of du Pont, a new field of ion-radical salts has been
opend. Our knowledge of organic solids has been improved very much
Is it still a new field even to-day ? I think it must be a main
problem to be discussed in this meeting. If it is, what do we mean by
a new field ? Probably it has two meanings as you know. One is a
virgin field of investigation to which we can apply principles or
methods that we have and used successfully in other cases. The
other means a field of investigation which demands quite a new
principle or a new theory. After nearly twenty years investigation, "anthracene" seems still requiring a theory for itself.

It seems me that there are two trends of approach to the problem. One is a way of physicists, I mean a stand-point of more

3

physical view, which demands most reliable informations and exact-
ness of parameters, but variety of materials is not needed and is
rather satisfied with anthracene, but it must be highly purified.
So much improved the purity of crystals in these ten years, it may
be one of important developments that have been made in these
years. The other is a way of chemists, I mean the chemical view
point, which makes the variety of compounds. TCNQ is a good example, it is still most attractive compound. The successful results
of crystal analysis of molecular compounds and ion-radical salts
made most important contribution in this field.

At any way, the studies of organic crystals have changed and
improved the chemical concept of molecule so much. **A molecule is**
not a closed system as considered before, but it is quite an open
system for electrons. I think this is the most important point
which we have realized from the studies of organic crystals. For
instance, a column of ion-radicals found in a crystal of TCNQ
salts can be considered as a giant linear molecule as a whole. In
this one-dimensional molecular array, there must be a chemical
binding between molecules, which is characterized with weak interaction but strongly specific properties.

Organic compounds mean much more than compounds made of carbon, it does mean originally the substances to make an organism.
I can say, we have now realized the meaning of organic compounds
why they are called organic compounds in its original meaning.

I sincerely hope that those participating in this meeting
will have something to offer regarding the study of the present
situation of organic solids and will contribute to its further development.

II. CHARGE CARRIERS

VOLTAGE DEPENDENCE OF UNIPOLAR EXCESS BULK CHARGE DENSITY IN ORGANIC INSULATORS

Martin Pope and William Weston

New York University, Radiation and Solid State Laboratory

4 Washington Place, New York, N.Y. 10003 U.S.A.

The steady flow of a unipolar current through a real insulator containing bulk trapping sites and provided with an injecting contact has been divided into several regimes[1]. At low voltage, there should be an ohmic regime in which bulk, thermal generation of carriers predominates. This is followed at higher voltages by a space charge limited current (SCLC) regime, during which traps are filled by the excess, injected carriers. With certain trap distributions, it is sometimes found that at a particular voltage (V_{TFL}) the current rises much more steeply with voltage than was the case for voltages less than V_{TFL}. This voltage is referred to as the traps filled limit (TFL) voltage and it provides a ready measure of the total trap density in the crystal. At sufficiently high voltages, the contact becomes depleted and the current tends toward saturation.

While there are examples in the literature that illustrate one or more of the regimes described in ref. 1, it has become evident that there may be entirely different physical mechanisms underlying the observed current voltage (J-V) response in real crystals. For example, in the work of Baessler et al.[2] on tetracene crystals of different thickness with different injecting electrodes, there is an Ohm's law region at low voltages that cannot be due to the thermal generation of bulk carriers because the current depends on the nature of the ohmic contact. Furthermore, in the work of Campos[3] on naphthalene, it is shown that the steeply rising portion of the J-V curve that usually provides the value of V_{TFL} and in addition, the density of trapping sites[1], cannot in fact be related to the TFL.

7

In view of some of the apparent discrepancies between the
simple SCLC theory[1] and the observed results, it is necessary to
determine the role played by the special conditions in real crystals. For example, the image force between injected charge and the
electrode has been neglected, although the image force can have a
profound effect on the J-V behavior[4]. Also, the concentration of
injected carriers at the injecting electrode is often assumed to be
practically infinite (see ref. 1 for discussion) while in fact, injecting electrodes often provide only a modest carrier concentration
at the electrodes, thereby affecting the J-V behavior appreciably[5].
In addition, diffusion currents are almost always neglected in theoretical treatments, but they can play a role in determining the J-V
behavior for applied voltages as high as 100V on a crystal about
200μ thick[6] if trapping is greater near the injecting electrode.
Thus, although it is usually assumed that trapping sites are uniform-
ly distributed in the bulk of the crystal, it is often the case that
there is a higher concentration of trapping sites near the crystal
boundaries and this can alter the observed J-V behavior[7]. It is
therefore a matter of importance to develop techniques that can provide the information necessary to specify the proper boundary conditions for theoretical treatments.

The problems mentioned above could be resolved if one knew the
spatial dependence of the electric field E inside the crystal.
Since this is often difficult to measure in thin insulating crystals,
an alternate approach would be to measure the spatial distribution
of charges in the crystal; attempts are now being made in our laboratory to do so. This paper, however, will deal with the voltage
dependence of the average excess charge density in the entire crystal.

To measure the excess charge density, we make use of the ability
of charge carriers to quench triplet excitons[8]. Triplet excitons
can undergo a bimolecular fusion process in anthracene (and other
crystals) to produce a singlet exciton that subsequently decays
with the emission of a delayed fluorescence (DF)[9]. The monomolecular decay rate of the DF can be shown to be twice that of the triplet exciton (β). The decay rate β can be increased by injecting
carriers into the crystal and the relation $\Delta\beta/\beta_o = \gamma_{Tt} \langle n_t \rangle \beta_o^{-1}$
can be derived where $\langle n_t \rangle$ is the average excess trapped carrier
density, γ_{Tt} is the bimolecular triplet-trapped charge interaction
rate constant, and $\Delta\beta = \beta_f - \beta_o$ where β_o, β_f are respectively the rate
constants for triplet exciton decay in the absence and presence of
the carrier density. It should be kept in mind that free carriers
will also quench triplet excitons, but in these experiments, trapped
carriers are present in excess. In the special case of SCLC[1],
$\langle n_t \rangle \approx \epsilon_o \epsilon V/eL^2$ where $\epsilon_o \epsilon$ is the dielectric permittivity of the
insulator, L is the insulator thickness, and e the electronic charge.

We have measured $\Delta\beta/\beta_0$ and the current density J as a function of voltage (V) for an anthracene crystal provided with a $Ce(SO_4)_2$ solution electrolytic contact[10]. The J-V dependence is shown in Fig. 1 and the $(\Delta\beta/\beta_0)$-V dependence is shown in Figs. 2 and 3. At low voltage (< 6V), the current in Fig. 1 shows evidence of a contribution from diffusion. At intermediate voltages, the current appears to be SCL and saturation commences at about 60V. At increasingly elevated voltages, the saturation current shows evidence of a high field effect.

As for the voltage dependence of the excess carrier density, it has previously been explained[6] that in the diffusion assisted regime, $\Delta\beta/\beta_0$ should be greater than would be the case if a purely SCLC flowed at that voltage; and at saturation, $\Delta\beta/\beta_0$ should be less than would be the case for a SCLC regime. In Fig. 2, a portion of the $(\Delta\beta/\beta_0)$-V dependence is shown; at voltages below 6V, there is evidence that $\Delta\beta/\beta_0$ lies above the line expected of a SCLC response, implying that diffusion is influencing this region. From the SCLC regime in Fig. 2, one calculates $\gamma_{Tt} \approx 5 \times 10^{-11}$ cm^3 s^{-1}. This value is about the same as that found for the triplet-triplet bimolecular fusion rate constant[11]. In many of the crystals we have studied, the measured value of γ_{Tt} was lower than 5×10^{-11} cm^3 s^{-1}, sometimes by more than an order of magnitude. In these crystals, the calculated density of carrier trapping sites (see ref. 1 for details) was comparable to or less than the estimated density of triplet exciton quenching sites n_q. n_q was estimated by assuming that β_0 in the virgin crystal was determined entirely by the quenching centers n_q. If $n_q > n_t$, then if an electron is trapped on an n_q site, it might be invisible as a quenching agent to the triplet exciton, since the n_q site would have quenched the triplet exciton anyway. This situation, together with an unequal distribution of trapped carriers, can account for the low values for γ_{Tt}. In Fig. 3, the marked drop of $\Delta\beta/\beta_0$ is indicative of a depleted contact and saturation. The saturation in Fig. 1 can be described by an equation of the form[12] $J = J_S + J_0 \exp (\beta V^{\frac{1}{2}} L^{-\frac{1}{2}})$ where J_S is the saturation current density at the onset of saturation, J_0 is a constant, and β is the Schottky constant $(e/kT) \times (e/4\pi\epsilon\epsilon_0)^{\frac{1}{2}}$. In the saturation regime, bulk space charge effects are not dominant, so one can assume for this regime that for any current J, $\langle n_f \rangle \approx e\mu v/LJ$ where $\langle n_f \rangle$ is the average free carrier density, and μ is the carrier mobility. Assuming a constant ratio of free to trapped carriers, one can normalize the expression for $\langle n_f \rangle$ (and hence $\langle n_t \rangle$) at V = 300V, whereupon all the other values of $\langle n_f \rangle$ become fixed, and as is shown in Fig. 3, the agreement is reasonable. There is reason to believe that the image force between a free carrier and an electrolyte solution should be very small because the high frequency or optical dielectric constant of water is the appropriate value to use for the image force calculations, and

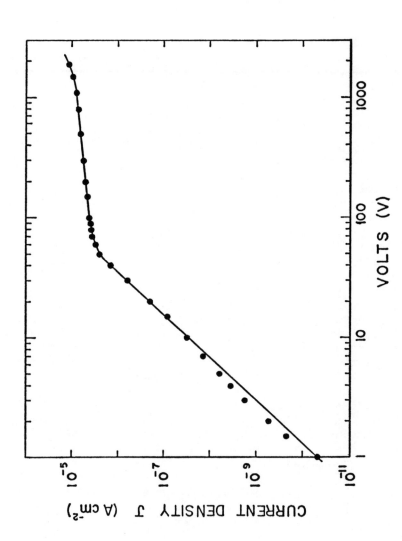

Fig. 1. Voltage dependence of hole current density through a 230μ thick anthracene crystal. Electrodes: $0.1M$ Ce^{4+} in $7.5M$ H_2SO_4 (anode) and H_2O (cathode). Electrode area is 1.6×10^{-2} cm^2.

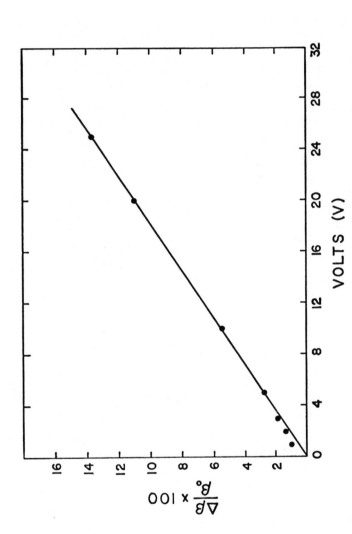

Fig. 2. Voltage dependence of average excess total charge density, $\langle n_{tot} \rangle \propto \Delta\beta/\beta_0$, in a 230μ thick anthracene crystal. Electrodes: 0.1M Ce^{4+} in 7.5M H_2SO_4 (anode) and H_2O (cathode). Measured value of β_0 is 53 s^{-1}; calculated value of $\gamma_{Tt} \approx 5 \times 10^{-11}$ cm^3 s^{-1}.

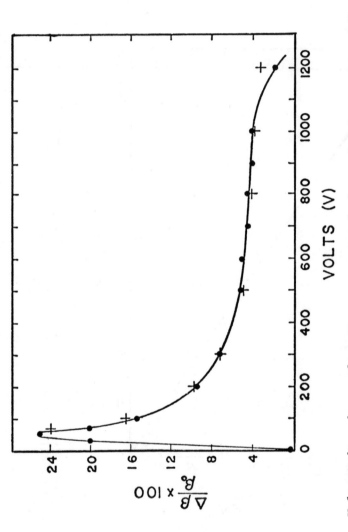

Fig. 3.   Voltage dependence of average excess total charge density, $n_{tot} \propto \Delta\beta/\beta_o$, in a 230μ thick anthracene crystal. Electrodes: 0.1M Ce4+ in 7.5M H_2SO_4 (anode) and H_2O (cathode). ● Experimental. ✛ Theoretical using equation $\langle n_f \rangle = JL/e\mu V = L \, (J_S \pm \frac{1}{2} \, J_o \, \exp \, (\beta L - \frac{1}{2}\sqrt{2}))/e\mu V$ normalized at V=300 volts with $\beta = 0.0083 \, cm^{\frac{1}{2}} V^{-\frac{1}{2}}$.

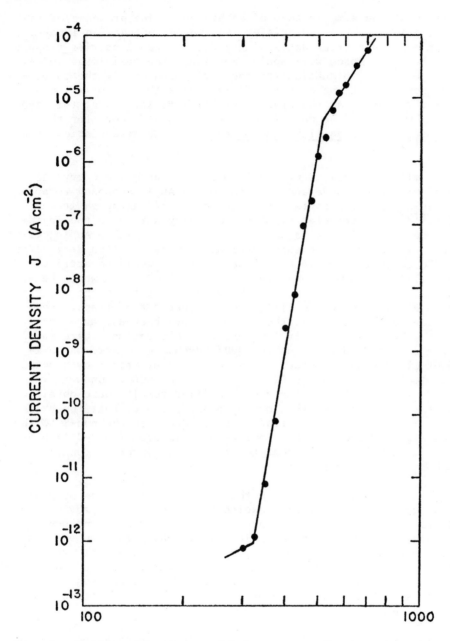

Fig. 4. Voltage dependence of hole current density through a 94μ thick anthracene crystal with a Au anode and Hg cathode. $\beta_o = 59$ s^{-1}.

this is about the same as that of anthracene. The presence of the
Schottky effect is thus surprising, and would indicate an important
role for trapped carrier detrapping process since a carrier trapped
within the image force well would experience the full image force.
Another possible explanation for the J-V dependence in the saturation region is that the electric field favors the dissociation of
hole-electron pairs at the crystal-electrolyte interface; this type
of dissociation would be described by the Onsager theory for the
field effect on the <u>equilibrium constant</u> for the dissociation of an
ion-pair[13,14].

A similar experiment was performed on an anthracene crystal
provided with a Au anode and a Hg cathode. Au is supposed to provide a hole injecting contact to anthracene[15], although it was
surmised[15] that the electric field may play a role in the injection process. In Fig. 4 is shown a J-V response with a Au injecting
contact. The response is quite typical of what is often attributed
to a SCLC current; the slope at lower voltages would be attributed
to the filling of a discrete trap level, the V^{32} slope would be attributed to a V_{TFL} response, and the V^8 slope at the higher voltages
would be attributed to the filling of another set of traps. However, despite the fact that there should have been a change in
$\Delta\beta/\beta_0$ at 400V of practically 100% under a SCLC regime, there was <u>no</u>
measurable change in $\Delta\beta/\beta_0$ up to 400V; beyond 400V there was evidence of electroluminescence. This is incontrovertible evidence
that at least up to 400V in Fig. 4 there is no SCLC. On another
crystal supplied with Au electrodes, similar results were obtained,
and in this crystal, it was shown by a $\Delta\beta/\beta_0$-V plot that the V^8
slope in Fig. 4 could be attributed to a SCLC. In the non-SCLC region of Fig. 4 the data fit a Fowler-Nordheim plot ($\ln J$-V^{-1}) pro-
viding additional evidence that the Au contact involves tunneling
through a surface barrier.

Summarizing, it can be seen that the lifetime of the DF produced by the fusion of triplet excitons can be used to provide a
qualitative and quantitative measure of the voltage dependence of
the excess carrier density in an insulator, which in turn opens an
entirely new window into the study of electrical conductivity in
insulators. Since the observable is an optical signal and the triplet exciton is unaffected by the electrical field, changes in the
charge density can be measured precisely and without reference to
internal or external field.

This work was supported by the National Science Foundation and
the Atomic Energy Commission. We are grateful to Mr. J. Burgos,
Dr. W. Mey and to Professor Dr. H.P. Kallmann for stimulating discussions.

REFERENCES

(1). M.A. Lampert and P. Mark, Current Injection in Solids,
    Academic Press, New York, 1970.
(2). H. Baessler, G. Hermann, N. Riehl, and G. Vaubel,
    J. Phys. Chem. Solids, 30, 1579 (1969).
(3). M. Campos, Mol. Cryst. and Liq. Cryst., 18, 105 (1972).
(4). H.P. Kallmann and M. Pope, J. Chem. Phys., 36, 2482 (1962).
(5). N. Sinharay and M. Meltzer, Solid State Electronics, 7,
    125 (1964).
(6). M. Pope and H.P. Kallmann, Israel Journ. Chem. 10, 269 (1972).
(7). A.I. Rozenthal and L.G. Paritskii, Sov. Phys. Semicond., 5,
    2100 (1972).
(8). W. Helfrich, Phys. Rev. Lett., 16, 401 (1966).

(9). H. Sternlicht, G.C. Nieman, and G.W. Robinson, J. Chem.
    Phys., 38, 1326 (1963).

(10). M. Pope and W. Weston, Mol. Cryst. and Liq. Cryst. (to be
    published).
(11). P. Avakian and R.E. Merrifield, Mol. Cryst., 5, 37 (1968).
(12). J.S. Bonham and L.E. Lyons, Aust. J. Chem., 26, 489 (1973).
(13). L. Onsager, J. Chem. Phys., 2, 599 (1934).
(14). N.E. Geacintov and M. Pope in Proc. 3rd Int. Photocond. Conf.
    Stanford 1969, ed. by Pell, Pergamon Press Ltd. 1970.
(15). W. Mehl and B. Funk, Phys. Lett., 25A, 364 (1967).

GEMINATE CHARGE-PAIR RECOMBINATION IN MOLECULAR CRYSTALS*

C.L. Braun and R.R. Chance

Department of Chemistry, Dartmouth College

Hanover, New Hampshire 03755

Intrinsic photoconductivity is now known in a number of molecular crystals. Beginning with the work of Castro and Hornig which established a threshold of 4.0 eV for anthracene,[1] values of 3.0 eV for naphthacene,[2] 3.8 eV for pyrene,[3] 4.9 eV for phenanthrene,[4] and 5.2 eV for naphthalene[5] have been measured. These values may be rationalized in at least semi-quantitative fashion using the ideas of Lyons[6] together with gas phase and crystal ionization potentials plus free molecule electron affinities.

With respect to the ionization mechanism, there is strong evidence that at least for anthracene crystals, photoionization proceeds via autoionization[7] of the neutral exciton states which dominate the absorption spectrum below 6 eV. Recent support for this conclusion comes from the observation that the excitation spectrum for intrinsic photoconduction is very similar to that for the radiationless process of singlet exciton fission.[8]

We have long been interested in formulating a description of the succeeding process; namely, that by which the geminate charge pair formed on ionization separates to produce a free hole and electron carrier. Several years ago we showed[9] that the electric field dependence of electron carrier quantum yields in anthracene agreed fairly well with predictions based on Onsager's theory of geminate recombination.[10] For fields below about 10^4V/cm, the theoretically predicted field

dependence of the free carrier yield is linear, with a
slope to intercept ratio of

$$S/I = \frac{e^3}{8\pi\varepsilon\varepsilon_0 k^2 T^2} \tag{1}$$

where ε is the high frequency dielectric constant and
ε_0(F/cm) is the permittivity of free space. We found
linear field dependence and experimentally S/I did vary
approximately as T^{-2}. However, the most critical prediction from Eq.(1) for anthracene which has a mean dielectric constant of 3.2 ± 0.1,[11] is that the value of
S/I at 298K should be (3.4 ± 0.1) x 10^{-5}cm/V. Instead
we found values about 50% larger. Later Geacintov and
Pope measured the anthracene photocurrent field dependence up to much higher fields than we had used but did
not resolve the question of whether experiment or theory
was deficient with respect to S/I values.[12]

However, our most recent experiments[13] agree essentially quantitatively with the predictions of Onsager
theory. Several improvements in our pulsed-light photocurrent measurements were necessary to achieve these results; the most important was to limit the number of
absorbed photons to about 2.6 x 10^{10} per flash. Curves
a and c of Figs. 1 and 2 are results for electron and
hole carriers, respectively. In both cases, the lowest
applied field datum is a result for a "virgin" crystal
which had not previously been excited in an applied electric field.

Intercepts and S/I values from Figs. 1a and 2c are
given in Table I. The experimental S/I values are seen
to be a bit smaller than the theoretically expected value
(3.4 ± 0.1 x 10^{-5}cm/V), a possible indication that charge
carrier escape is most probable along directions of high
dielectric constant.

Table I
Experimental Slope to Intercept Ratios (S/I) and
Zero Field Quantum Yields (ϕ_0).

Free Carrier	S/I (10^{-5}cm/Volt)	ϕ_0 x 10^5 [b]
Electrons (Fig. 1a)	3.21 ± 0.12	8.75
Holes (Fig. 2c)	3.02 ± 0.08	9.16[a]

[a] The extrinsic contribution to the hole photocarrier
quantum yields was estimated to be less than 3%.

[b] The absolute error in the ϕ_0 values is estimated
to be about ±25%.

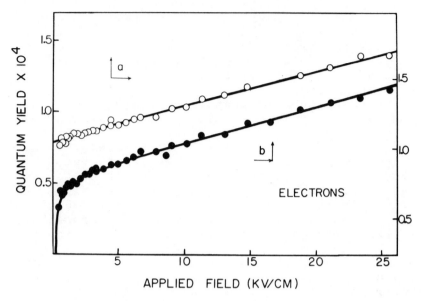

Figure 1.   Electron Quantum Yields at 255 nm.   Curve a
is for a virgin crystal sample.   Curve b illustrates the
reduced low-field yield which occurs when trapped holes
are present in the excited volume of the crystal.

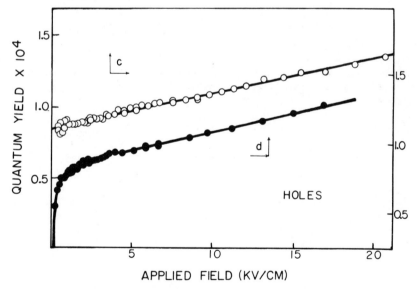

Figure 2.   Hole Quantum Yields at 255 nm.   Curve c is
for a virgin sample while in curve d trapped electrons
reduce the low-field free hole yield.

Reduced yields at low fields are observed in the Fig. 1b and 3d experiments which were performed after the Fig. 1a and 2c experiments, respectively. A quantitative interpretation of the Fig. 1b and 2d results can be made in terms of the recombination of free carriers with the opposite-signed, trapped charges left in or near the excitation volume during the earlier experiments. Formulation of appropriate free-trapped recombination kinetics[13] leads to an expression which was fit to the Fig. 1b and 3d data yielding the solid lines shown. The values of the second order rate constants for free-trapped recombination which were extracted from the least squares fits shown are 6.2 and 6.6 x 10^{-7} cm^3/sec for free electrons and free holes, respectively. These values are within 10% of the values predicted by Langevin theory of second-order recombination--agreement that is well within the expected accuracy of our kinetic model.

Several additional inferences can be drawn from the fact that there is nothing anomalous about the lowest-field portions of the Fig. 1a and 2c data. The free carriers observed in these experiments were produced initially within an absorption depth of about 10^{-5} cm from the crystal surface. At the lowest fields employed--a few hundred volts per centimeter--drift of the carriers in the applied field is slow enough that most of the free carriers have an opportunity to diffuse to the crystal surface. If such diffusion is energetically allowed, then the absence of any observed effects rules out both surface recombination and surface trapping as significant. Thus a clean anthracene surface does not seem to interfere in any observable way with the generation of free carriers.

Further support for the applicability of Onsager theory comes from the results of the lower-temperature, field dependence studies found in Fig. 3. Points on the lower curve in Fig. 3 give virgin crystal results for the temperature dependence of S/I. The solid curve is a plot of the theoretically predicted temperature dependence of S/I with ε of Eq. (1) adjusted upwards by 10% consistent with the room temperature datum. Thus the T^{-2} dependence of S/I expected from Onsager theory is found to fit the experimental results exactly. The upper curve in Fig. 3 shows the increasingly large deviations from theoretically-expected S/I values that are found in non-virgin crystals as the temperature is lowered. This striking effect is quantitatively explicable in terms of increased free-trapped recombination coefficients arising

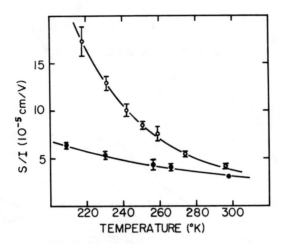

Figure 3. Experimental S/I values at 255 nm as a Function of Temperature. ● - virgin crystal results; O - non-virgin crystal results. The lower curve is the Onsager prediction assuming a room temperature S/I of 3.1×10^{-5} cm/volt.

from increased free carrier mobilities at low temperatures.[14]

Thus, for "virgin" crystals, Onsager theory appears to be essentially quantitative, while for crystals containing trapped charges, the observed free carrier yield reductions can be quantitatively understood in terms of free-trapped recombination.

The absence of adjustable parameters in the low-field limit of Onsager theory makes these experiments a very sensitive test of the applicability of the theory to photogeneration of charge carriers. The better-known, Poole-Frenkel theory is a highly-simplified attempt to deal theoretically with the same experimental situation of external-field-assisted carrier escape from a coulomb well. The top curve in Fig. 4 shows the steep field dependence predicted from the conventional, one-dimensional form of Poole-Frenkel theory.[15] Whereas for

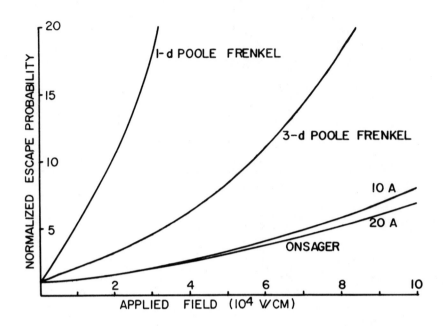

Figure 4. Comparison of Onsager Theory with Poole-Frenkel Theory. The upper curve is a plot of exp α versus applied field E where α is defined as $[e^3E/(\pi\varepsilon\varepsilon_0 k^2 T^2)]^{1/2}$. The plot gives the relative field dependence predicted by one-dimensional Poole-Frenkel theory. The middle curve shows the results of a pseudo three-dimensional form of Poole-Frenkel theory in which the relative effect of the applied field increases as $1/2 + [1 + (\alpha-1)\exp\alpha]/\alpha^2$. The lower curves give the normalized field dependence predicted by Onsager theory for r_0 values of 10 and 20A. Larger r_0 values would give even less upward curvature.

ε = 3.2 and T = 300K, Onsager theory predicts a modest 34% increase in free carrier yield between zero and 1×10^4 V/cm, Poole-Frenkel theory predicts a 516% increase. The origin of this striking difference is that in the one-dimensional, Poole-Frenkel theory, the geminate charge pair is forced to separate only in the maximally field-assisted direction. In contrast, the exact treatment of the classical problem afforded by Onsager theory allows diffusion of the geminate pair on the combined coulombic-applied potential surface and separation in any direction.

Thus Poole-Frenkel theory strongly overestimates the effect of the applied field by assuming that the field defines a unique configuration coordinate for geminate pair separation.

A modified form of Poole-Frenkel theory has been proposed by Jonscher[16] and by Hartke[17] and its predictions for our case are illustrated by the middle curve in Fig. 4. In this theory the geminate pair is allowed to separate along any radius vector within 90 degrees of the applied field axis and the barrier lowering by the applied field is reduced appropriately for the off-axial directions. This procedure obviously reduces the effect of the applied field and this pseudo three-dimensional Poole-Frenkel theory thus agrees better with Onsager theory and with experiment. However, it is still fatally flawed by implicitly forbidding each carrier pair to move except along some particular radius vector. Thus there is clear evidence that Onsager theory is the appropriate theory for use in interpreting the role of geminate recombination in free carrier production in anthracene single crystals. Furthermore, applicability to photogeneration in a wide range of other materials seems guaranteed by the nature of the basic theory which requires only that classical diffusion adequately describe the geminate-pair motion.

Finally, we are confident that with the theory of free carrier production in low-mobility solids seemingly on firm ground, extensions of these studies to high fields, various photon energies and various temperatures will allow real insight into the general question of low-energy electron scattering in these materials.

References

*This work was supported in part by the National Science Foundation.
1. G. Castro and J.F. Hornig, J. Chem. Phys. 42, 1459 (1965).
2. N. Geacintov, M. Pope and H. Kallmann, J. Chem. Phys. 45, 2639 (1966).
3. (a) R.F. Chaiken and D.R. Kearns, J. Chem. Phys. 49, 2846 (1968). (b) G. Castro, IBM J. Res. Dev. 15, 27 (1971).
4. R.G. Williams and B.A. Lowry, J. Chem. Phys. 56, 5736 (1972).
5. J. Aihara and C.L. Braun (unpublished).
6. F. Gutmann and L.E. Lyons, Organic Semiconductors (John Wiley & Sons, Inc., New York, 1967) Chap. 6.

7. N. Geacintov and M. Pope, J. Chem. Phys. <u>47</u>, 1194 (1967).

8. G. Klein, R. Voltz, and M. Schott, Chem. Phys. Lett. <u>19</u>, 391 (1973).

9. (a) R.H. Batt, C.L. Braun, and J.F. Hornig, J. Chem. Phys. <u>49</u>, 1967 (1968). (b) R.H. Batt, C.L. Braun, and J.F. Hornig, Applied Optics, Supplement 3, 20 (1969).

10. L. Onsager, Phys. Rev. <u>54</u>, 554 (1938).

11. I. Nakada, J. Phys. Soc. Japan, <u>17</u>, 113 (1962).

12. N.E. Geacintov and M. Pope, Proc., Intern. Photo-conductivity Conf. 3rd, Stanford, Ca., 1969, p. 289 (1971).

13. R. Chance and C.L. Braun, J. Chem. Phys. <u>59</u>, 2269 (1973).

14. R. Chance, J. Electrochemical Society, in press.

15. J. Frenkel, Phys. Revs. <u>54</u>, 657 (1938); R.M. Hill, Phil. Mag. <u>23</u>, 59 (1971).

16. A.K. Jonscher, Thin Solid Films <u>1</u>, 213 (1967).

17. J.L. Hartke, J. Appl. Phys. <u>39</u>, 4871 (1968).

VARIOUS DETRAPPING PROCESSES IN ANTHRAQUINONE-DOPED

ANTHRACENE CRYSTAL

Uichi Itoh and Katsuji Takeishi

Electrotechnical Laboratory, Tanashi, Tokyo

The photo-carrier generation process by optical detrapping effects has been observed in anthracene crystals provided with hole[1] and electron[2] injecting contacts. It has been reported that there were the direct release of trapped carriers from traps about 1 eV to conduction state and the indirect release due to the interaction of the trapped carriers and two types of excitons, that is, triplet and singlet excitons. On the other hand, it is expected that there are deeper traps than 1 eV in impurity-doped anthracene crystals. In fact, it was found that free carriers in electrical conduction were generated thermally from such deep traps in doped anthracene crystals.[3] In this note, various detrapping processes which include the direct release from the deeper traps than 1 eV are investigated in anthraquinone-doped anthracene crystal. Further, it is reported that the enhancement of the photocurrent by the optical detrapping effects is observed in the crystal provided with non-injecting contacts, when the traps have been charged with carriers before measurements.[4,5]

The anthraquinone-doped crystal was grown by a Bridgman type furnace. The content of the anthraquinone was determined to be about 170 ppm by a gas chromatography. The crystal was sandwiched with two SnO_2 coated quartz plates which were used as transparent electrodes. The crystal was illuminated with monochromatic light through one of the electrodes. The steady-state photocurrent was measured with a vibrating reed electrometer, and was corrected for incident photon numbers.

The excitation spectra of hole and electron photocurrent are shown in Fig. 1. The structures of the spectra vary with the applied electric field. In the low field, the peak of the photo-

currents are observed in the region of the wavelength from 420 nm
to 440 nm.   In the high field, this peak hides behind the tail of
the large peak which appears at 395 nm.   However, when many carriers
were charged in the deep traps before the measurement of the photocurrent, the peak of the photocurrent appears from 420 nm to 440 nm
even in the high field, as shown by the dotted line in Fig. 1.   The
increase of the photocurrent after the charging indicates that the
carrier generation due to the detrapping effect is dominant.   Therefore, the variation of the spectra with the applied field is understood as the competition between the following two carrier generation
processes.   In the low field, most carriers are generated by the
detrapping effect and in the high field the carrier generation by
the exciton-surface interaction becomes dominant.

The treatment for the charging of the deep traps with carriers
was carried out by the illumination of the light at 395 nm under the
application of the high field for about one hour.   The carriers
generated in the surface were injected into the crystal and some of
them were trapped in the course of transit process.   The polarity
of the trapped carriers was prescribed by the polarity of the illuminated electrodes.   After the charging treatment many trapped

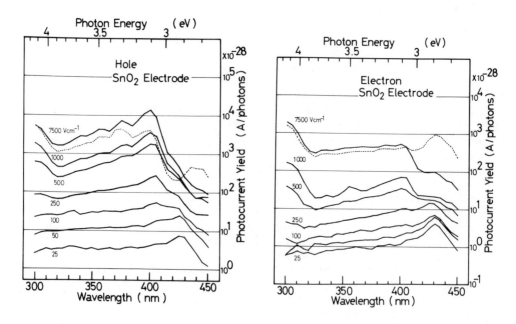

Figure 1.   Field dependence of excitation spectra for hole and
electron photocurrents.   The dotted line shows the photocurrents
in 7500 V/cm after the charging treatment.

carriers have been left in the crystal.   The photocurrent enhanced
by the detrapping effect is observed in the long wavelength region,
since the light in this region penetrates into the crystal.   The
variation of the excitation spectra of the photocurrent in the crystal subjected to the charging treatment are shown in Fig. 2.   In
the long wavelength region, it is clear that the photocurrent is
enhanced by the optical detrapping effects when the charging treatment was carried out before measurement.   The structures of the
photocurrent spectra show that there are different detrapping processes for the trapped holes and the trapped electrons.   The spec-

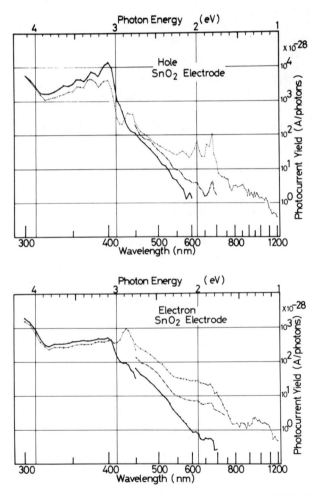

Figure 2.   Excitation spectra of photocurrents in 7500 V/cm.
⋯⋯⋯⋯⋯: when the charging treatment was carried out before
measurement, —·—·—·—: when some of the trapped carriers were
released, ————: when the trapped carriers were cleaned out.

trum of the hole photocurrent enhanced by the detrapping for the
trapped holes were composed of three region, that is, below 1.5 eV,
1.5 eV-2.5 eV and 2.5 eV-2.9 eV.   The spectrum below 1.5 eV is
interpreted by the direct optical release of the trapped holes from
hole-trapping level at about 1 eV as discussed by Adolph et al.[1]
The structure of the spectrum from 1.5 eV to 2.5 eV is the same as
that of the singlet-triplet absorption spectrum in anthracene crystal.[6]   Therefore, it is explained by the indirect release of the
trapped holes by the interaction of them and triplet excitons.
Further, the peak at about 2.8 eV is considered to be due to the
interaction of the trapped holes and singlet excitons.

The enhancement of the electron photocurrent due to the interaction of trapped electrons and two types of excitons was also observed.   However, the spectrum of the electron photocurrent enhanced
by the detrapping effects indicates that the direct optical release
 of the trapped electrons is dominant process rather than indirect
release by the interaction of excitons-trapped electrons.   As
discussed by Many et al. [2] the direct optical release from a
trapping level takes form of shoulders in the spectrum of the enhanced photocurrent.   Three shoulders were observed in the spectrum.
It suggests that three electron trapping levels exist.   From the
rising points of the shoulders, the values of three trapping levels
for electron are estimated.   Those are 1 eV, 1.6 eV and 2.3 eV.
It is noticed that the values of the two deeper electron trapping
levels coincide with those of two levels estimated from the temper-
ture dependence of the dark conduction.   The values of the activation energy of the dark conduction in the anthraquinone-doped
anthracene crystal [3] were 1.6 eV in the low temperature region
and 2.4 eV in the high temperature region.   Their values are summarized in Table 1.

Table 1.   Energy depth of hole and electron trapping levels which
were estimated from the spectra of the enhanced photocurrent and
from temperature dependence of conductivity.

Energy Depth		Thermal Activiation Energy of Conductivity
hole trap	electron trap	$\sigma = \sigma_0 \exp(-E/kT)$
1 eV	1 eV	
	1.6 eV	1.6 eV (380~390 K)
	2.3 eV	2.4 eV (390~420 K)

Finally, the rate constant of the interaction of triplet exciton and trapped hole was roughly estimated from the time dependence of the decrease of the hole photocurrent from 1.5 eV to 2.5 eV.   As shown in Fig. 3 the photocurrent enhanced by the detrapping effect showed the peak value immediately after the charging treatment and decreased gradually as the decrease of the number of the trapped carriers which were the source of the carrier generation by the detrapping effect.   If all free holes are generated only by the interaction of triplet exciton-trapped hole, the following expressions,

$$n_{free} \propto n_{trap} \qquad (1)$$

$$\frac{d\, n_{trap}}{d\, t} = -\, k\, n_T\, n_{trap} \qquad (2)$$

are held where n_{free} is the density of the free holes, n_{trap} is the density of the trapped holes which have been left without interaction with triplet excitons, k is the rate constant and n is the density of the triplet excitons. n_{free} and n_T are estimated from the values of the hole photocurrent and the singlet-triplet absorption coefficient of anthracene crystal, respectively.   k calculated from the slope of the decrease of the hole photocurrent is 5.5 x 10^{-11} cm^{3}sec^{-1}.   This value is about the same order as that obtained by Ern et al.[7]   They estimated the value to be 0.7 x 10^{-11} cm^{3}sec^{-1} from the measurement of the lifetime of the delayed fluorescence in anthracene crystal with hole injecting contact.

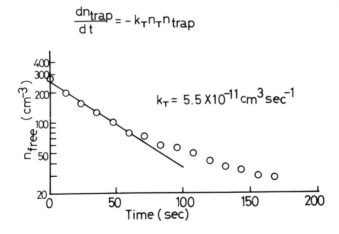

Figure 3.   Time dependence of hole photocurrent after charging treatment.

Table 2

	Injecting Contacts	Hole	Electron
Adolph et al. (1964)	hole	1.1 eV S_1 ,T_1 – trapped hole	
Many et al. (1969)	electron		0.95 eV S_1 ,T_1 – trapped electron
Ours (1973)	non	1.0 eV S_1 ,T_1 – trapped hole	1.0, 1.6, 2.3 eV S_1 ,T_1 – trapped electron

In conclusion, it is pointed out that various detrapping processes play important role in the carrier generation processes of the dark conduction and photoconduction in anthracene crystal. The various detrapping processes are summarized in Table 2.

References

[1]   J.Adolph, E.Baldinger and I.Granacher; Phys. Letters 8 (1964) 224.
[2]   A.Many, J.Levinson and I.Teucher; Mol. Cryst. 5 (1969) 273.
[3]   U.Itoh, K. Takeishi and H.Anzai; J.Phys.Soc.Japan 35 (1973) 810.
[4]   U.Itoh; J.Phys.Soc.Japan 35 (1973) 514.
[5]   U.Itoh and H.Anzai; submitted to J.Phys.Soc.Japan.
[6]   P.Avakian, E.Abramson,R.G.Kepler and J.Caris; J.Chem.Phys. 39 (1963) 1127.
[7]   V.Ern, H. Bonchriha, J.Fourny and G.Delacote; Solid state Comm. 9 (1971) 1201.

UV EXCITATION OF ANTHRACENE

Mitsuo Kawabe, Kohzoh Masuda and Susumu Namba

Faculty of Engineering Science, Osaka University

Toyonaka,Osaka, Japan

In order to study the exciton diffusion or energy transfer in anthracene crystals, pulse excitations are often used. Some investigators used UV light to avoid the reabsorption.

In this paper, the change in the fluorescence decay time of anthracene excited by a nitrogen gas laser is discussed from the view point of 1) the effect of reabsorption and 2) the high density excitation.

I THE EFFECT OF REABSORPTION

The fluorescence decay time of pure anthracene is affected by reabsorption, especially above liquid nitrogen temperature. Birks has studied the effect of reabsorption and expressed this effect as follows[1],

$$\tau = \frac{\tau_0}{1 - aq}$$

where τ_0 is the molecular decay time defined by the fluorescence decay time where the reabsorption is negligible, a is the fraction of primary-fluorescence photons reabsorbed in the crystal and q is the fluorescence quantum efficiency of the crystal. However, since the magnitude of the reabsorption depends on the spectral overlap of the emission and absorption, the fluorescence decay time differs for different wavelength. Because of the large absorption coefficient, the fluorescence of short wavelength comes from near the surface of the crystal; thus the decay times are not affected very much by reabsorption, while the intensity of the fluorescence is very small. On the other hand, the long-wavelength fluorescence is transparent in the crystal; therefore, the light which is emitted from the bulk of the crystal can be detected.

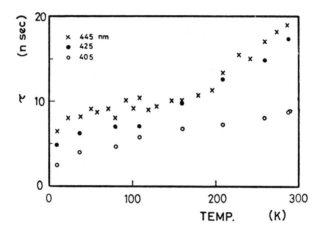

Fig. 1   Temperature dependence of the front-surface
          fluorescence decay time.

Figure 1 shows the temperature dependence of the front-surface
fluorescence decay time at different wavelengths which correspond
to the three peaks arising from vibrational modes in anthracene.
The qualitative explanation is as follows.   Two kinds of fluorescence can be received by the detector.   The one I_d is emitted directly from initially excited excitons without suffering reabsorption and the other I_r is emitted after reabsorption and emission
cycles.   The effect of reabsorption on the decay time of the observed fluorescence depends on I_r/I_d.   The intensity of the direct
fluorescence is experssed as

$$I_d = AS \int_0^\infty n_o e^{-(\alpha_o + \alpha_1)x} dx = \frac{ASn_o}{\alpha_o + \alpha_1} \qquad (1)$$

where A is a constant, S is the illuminated area, x is the distance
from the surface, $n_o \exp(-\alpha_o x)$ is the distribution of excitons excited by a nitrogen gas laser and α_1 is the absorption coefficient at
the detected wavelength.   As for I_r, photons emitted from Sdx are
partly absorbed and the amount absorbed in the volume Sdx' at x' is

$$dI = BSn(x)\alpha(\omega)e^{-\alpha(\omega)|x-x'|} dx \, dx' \, g(\omega)d\omega. \qquad (2)$$

B is assumed to be constant for the case that the source of the
luminescence is a plane, the area of which is enough large compared
with excited depth.   In this case B is slightly smaller than 0.5
because most of the fluorescence is emitted equally to both sides
of the plane.   In the case of point-source luminescence, on the

other hand, B is proportional to the inverse square of the distance. $\alpha(\omega)$ is the absorption coefficient of anthracene and $g(\omega)$ is the spectral distribution of the fluorescence. Assuming that the fluorescence efficiency is unity, the reemitted fluorescence after suffering one reabsorption cycle is

$$I_{r,1} = \frac{ASn_o}{\alpha_o + \alpha_1} B \int_0^\infty \left\{ \frac{\alpha(\omega)}{\alpha_o + \alpha(\omega)} + \frac{\alpha(\omega)}{\alpha_1 + \alpha(\omega)} + \frac{0.9(\alpha_o + \alpha_1)\alpha(\omega)}{(\alpha_o + \alpha(\omega))(\alpha_1 + \alpha(\omega))} \right\} g(\omega) d\omega.$$

$$\text{-------(3)}$$

The third term in the integral is due to the light reflected back into the crystal from the front-inner surface of the crystal. In this case it is assumed that 90 % of the light which is incident upon the inner surface is reflected back into the crystal[2] and the reflected light from the back surface is neglected because most of the short-wavelength fluorescence is reabsorbed before reaching the back surface of the sample. Using the room temperature values of α_o[3], α_1, $\alpha(\omega)$[4] and $g(\omega)$, the integral of equation (3) is calculated to be 0.67 for the fluorescence of 405 nm. The total fluorescence I_r which suffers more than one reabsorption is

$$I_r = \sum_i I_{r,i} = \frac{ASn_o}{\alpha_o + \alpha_1} \frac{0.67B}{1 - 0.67B} \quad . \qquad (4)$$

In the case of fluorescence of the longer wavelength (445 nm), which is transparent in the crystal, the light can come from both the surface and deep region of the crystal without suffering reabsorption. The fluorescence intensity without suffering reabsorption is obtained from equation (1) putting $\alpha_1 = 0$: namely,

$$I_d = \frac{ASn_o}{\alpha_o} \qquad (5)$$

The fluorescence intensity $I_{r,1}$, in this case, is calculated in the same manner as for the short wavelength, and is expressed as fllows,

$$I_{r,1} = \frac{ASn_o}{\alpha_o} B \int_0^\infty \left\{ \frac{\alpha(\omega)}{\alpha_o + \alpha(\omega)} + 1 + \frac{0.9\alpha_o}{\alpha_o + \alpha(\omega)} \right\} g(\omega) d\omega = \frac{ASn_o \times 1.99B}{\alpha_o} \quad .$$

$$\text{------ (6)}$$

The total reabsorption fluorescence is, therefore,

$$I_r = \frac{ASn_o}{\alpha_o} \frac{1.99B}{1 - 1.99B} \qquad (7)$$

For the short wavelength, the intensity ratio of the reabsorbed light I_r to that of the direct light I_d is 0.37, on the assumption that B is 0.4.   For the long wavelength, the ratio is 4.0.
From these calculations, the short-wavelength fluorescence suffers
little reabsorption; on the other hand, most of the long-wavelength
consists of reabsorbed light.   The previously proposed decay times
at room temperature lie between 5 and 11 nsec.   The smaller value
is the result of surface fluorescence with theoretical correction
for reabsorption[5], and the larger one is the result extrapolated
to zero thickness when changing the sample thickness[6].   The decay
time of the short wavelength in Fig. 1 is a directly measured value
with little reabsorption.   The temperature dependence is similar
to the results of Takahashi *et al.*[6] but exhibits a smaller difference between high and low temperature.   The effect of reabsorption
decreases with decreasing temperature but even at liquid helium
temperature the effect still remains.

II   THE EFFECT OF HIGH DENSITY EXCITATION

The effects of high density excitation on anthracene fluorescence are i) decrease in decay time, ii) saturation of the fluorescence with increasing excitation, iii) increase in the absorption coefficient and iv) change in the fluorescence spectrum.
The decrease in decay time is due to singlet exciton-exciton interaction.   The saturation of the fluorescence intensity is due to
energy loss used to excite one exciton to higher energy level by
two exciton collision.   Figure 2 shows the fluorescence intensity
vs excitation intensity.   The total intensity (o) and the peak
intensity (•) show different behavior.

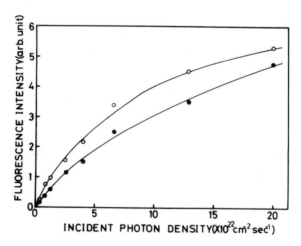

Fig. 2   Change in the fluorescence intensity with excitation
    intensity.   o --- total fluorescence
              • --- peak fluorescence

The singlet exciton-exciton interaction coefficient is estimated from the fluorescence decay curve using following equation,[7]

$$\frac{dn}{dt} = -\beta n - \frac{\gamma}{2} n^2 \tag{8}$$

$$\frac{1}{n} = (\frac{\gamma}{2\beta} + \frac{1}{n_0}) e^{\beta t} - \frac{\gamma}{2\beta} \tag{9}$$

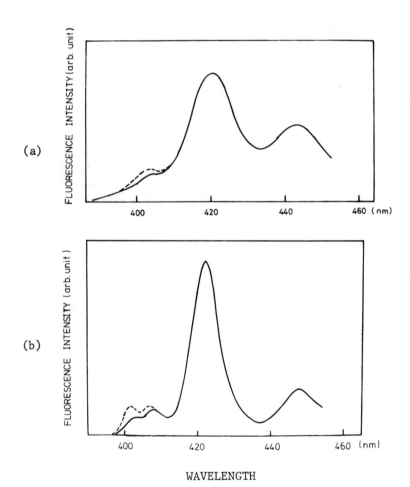

WAVELENGTH

Fig. 3 Change in the fluorescence spectrum at weak and strong excitation intensity.
   (a)   at room temperature, --- $\sim 10^{21}$ (photons/cm^2), —— $\sim 10^{23}$
   (b)   at liquid nitrogen temperature, --- $\sim 10^{20}$, —— $\sim 10^{23}$

where n is exciton density, β is monomolecular decay probability
and γ in bimolecular decay probability of excitons. As shown in
Table 1, estimated values of γ depend on the incident power and
the fluorescence wavelength.

Table 1 γ at various excitation densities

$\gamma(cm^3 \cdot sec^{-1})$	Wavelength (nm)	excitation (photons/$cm^2 \cdot sec$)	$1/\beta$ (ns)
1.06×10^{-9}	405	1.4×10^{21}	8.6
1.28×10^{-9}	445	1.6×10^{21}	19
2.2×10^{-9}	405	1.3×10^{22}	8.6
2.4×10^{-10}	445	1.3×10^{22}	19
1.7×10^{-11}	445	4.8×10^{23}	19

At the low excitation density ($\sim 10^{21}$ photons/$cm^2 \cdot sec$), γ is
about 10^{-9} $cm^3 sec^{-1}$, but at high excitation ($\sim 10^{22}$ photons/$cm^2 \cdot sec$),
γ at 445 nm is smaller than that of 405 nm. This is because of
the reabsorption. At the higher excitation density γ becomes order
of 10^{-11}. This is because of over estimation of the exciton
density in addition to reabsorption.

Figure 3 in the next page shows the change in the fluorescence
spectrum with excitation intensity at room temperature (a) and
liquid nitrogen temperature (b). Under strong excitation (solid
line in the figure) at both temperatures the short-wavelength
peak decreases just as the effect of reabsorption increases. These
results were not observed when the fluorescence was detected from
back surface because of masking the short-wavelength fluorescence
by reabsorption. The reason of increase in reabsorption under
strong excitation is not known. The increase in absorption
coefficient may be the reason of this phenomenon.

REFERENCES

1) J.B. Birks, Proc. Phys. Soc., 79 494 (1962).
2) B.J. Mulder, Phillips Res. Rep. Suppl., No.4 (1968).
3) H.C. Wolf, Z. Naturf. 13a 414 (1958).
4) I. Nakada, J. Phys. Soc. Japan 20, 346 (1965).
5) L.M. Logan, I.H. Munro, D.F. Williams and F.R. Lipsett, Molecular
    Luminescence ed. E.C. Lim (Benjamin, New York 1969) p.773.
6) Y. Takahashi and M. Tomura, J. Phys. Soc. Japan 31 1100 (1971).
7) A. Bergman, M. Levine and J. Jortner, Phys. Rev. Letters 18
    593 (1967).

PHOTOCONDUCTION AND EMISSION OF PHTHALOCYANINE IN THE NEAR INFRARED

Katsumi YOSHINO, Keiichi KANETO, Yoshio INUISHI

Faculty of Engineering, Osaka Univ. Yamada-Kami

Suita, Osaka, Japan

The emission spectra of phthalocyanine single crystals and evaporated thin films have been investigated in the near infrared region.   These emission spectra are discussed in connection with S-T interaction caused by central metal atoms.
Dependence of induced current on ruby and glass laser intensity shows that the triplet state plays an important role in the carrier generation in the near infrared region.

1. INTRODUCTION

The phtoconductive response of phthalocyanine in the near infrared attracted attention of many workers[1][2] in terms of the singlet-triplet absorption and defect states.   However, few papers have been reported on the direct evidence for such processes as phosphorescence emission in the phthalocyanine crystal.

In the present paper we will report the emission spectra of H_2Pc (metal free phthalocyanine), CuPc (copper phthalocyanine), ZnPc, PtPc, SnPc, NiPc and CoPc single crystals at temperature range from $1.8°K \sim 300°K$ in connection with S-T interaction caused by central metal atoms.   The emission spectra of evaporated thin film of the above mentioned phthalocyanine and PdPc are also investigated.   Dependence of induced photocurrent of ZnPc on ruby and glass laser light intensity are studied to clarify the carrier generation mechanisms in the near infrared region.

2. EXPERIMENTALS

H_2Pc, CuPc, CoPc, ZnPc, NiPc, PtPc, PdPc and SnPc were synthesized following   the method of Linstead[3] or by heating the phthalo-

nitrile and an appropriate metal in an evacuated pyrex glass tube
at 250~300°C for several hours.   To advance to the crystal
growth, these materials were purified by sublimation five
times.   The single crystal was grown by sublimation in
nitrogen carrier gas under reduced pressure of 2~5 mmHg.
The thin film was obtained by evaporating an appropriate sample on
a pyrex glass plate in vacuum under 10^{-6} mmHg.   In the case of
PdPc, a single crystal of sufficient size was hard to obtain, so
that only the evaporated thin film was studied.

The exciting light was supplied from a 1 KW Xe arc lamp and
the emission light was detected by a selected RCA 7102 photomulti-
plier cooled at Liq.N$_2$ temperature after passing through a Bausch
Lomb grating monochromator.   Q-switched ruby and glass lasers
of 30 n sec pulse width were used as the pulse light source for the
photoconductivity measurement.   A 650 W tungsten lamp was also
used as a light source for the photocondution spectra measurment.

3. RESULTS AND DISCUSSIONS

The photocondutive response of CuPc and H$_2$Pc in the near infrared as shown in Fig.1 revealed that the photoconductive response
at 1.09 μ were observed only in CuPc, but not in H$_2$Pc.
On the other hand, in H$_2$Pc the fluorescence emission near 780 mμ
was observed as shown in Fig.2 but not around 1 μ.
In CuPc however, the emission was observed in the near infrared at
1.12 μ but not in the range between 700~900 mμ.   These facts
may be explained as follows.   The magnitude of spin-orbit coupling of CuPc is larger than that of H$_2$Pc because of paramagnetic
and heavy metal effects.   Accordingly in CuPc with large spin-

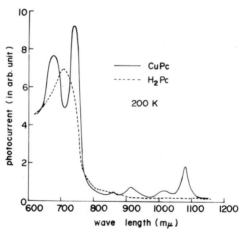

Fig.1   Photoconduction spectra of H$_2$Pc and CuPc
        single crystals in the near infrared

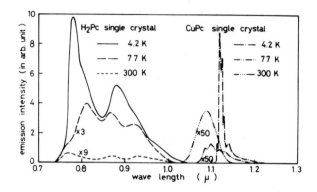

Fig.2   Emission spectra of H$_2$Pc and CuPc single crystals

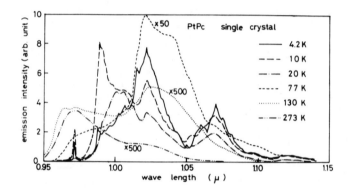

Fig.3   Emission spectra of PtPc single crystal

orbit coupling, the strong intersystem crossing quenches the fluorescence at 700~900 mμ and induces the strong phosphorescence at near 1.12 μ.   In H$_2$Pc with small S-T interaction, the strong fluorescence at 780 mμ was observed without the phosphorescence at around 1 μ.   These facts are consistent with the above mentioned data of photoconduction spectra.   Namely the observed response in the near infrared region in only CuPc at 1.09 μ is due to the S-T absorption caused by large spin-orbit coupling.

In ZnPc single crystal both the fluorescence (700 mμ~ 1 μ) and phosphorescence (1.15μ at 4.2°K) were observed.   This fact is reasonable because the magnitude of the spin-orbit coupling of ZnPc is considered to be the intermediate value between those of CuPc and H$_2$Pc.

In PdPc (thermal treated thin film), both the fluorescence

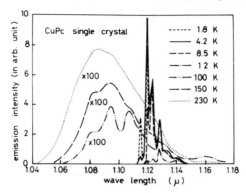

Fig.4 Temperature dependence of phosphorescence
       spectra of CuPc single crystal

around 820 mμ and phosphorescence around 990 mμ were observed.
In PtPc which contains very heavy metal, only the phosphorescence
was observed as shown in Fig.3.
     On the contrary to the cases of the above mentioned phthalo-
cyanines, no emission was observed in NiPc and CoPc single crystals
for all temperature ranges.   These facts may be explained in a similar was as experiments performed in porphyrin by Becker et al.[4]
as follows.   The excited states of the central metal d electrons
should lie below the triplet states of the ligands for these
samples and the radiationless decay via these excited d state
should quench the emission in these samples.
     Fig.4 show the temperature dependence of the phosphorescence
emission of CuPc single crystal.   The sharp emission peaks at
1.12 μ observed at Liq.He temperature begin to decrease markedly with
rising temperature from the shorter wave length side and the
intensity of the emission above 20°K was weaker than that at 4.2°K
by about two orders.   By raising temperature further, new emission line appears at 1.09 μ and increases with the activation
energy of about 20 meV which coincides nearly with the energy difference between peaks at 1.09 μ and 1.12 μ.   The decay time of
this emission was also observed by the Q-switched ruby laser excitation.   The decay time of the emission at 1.12 μ was nearly
2∼3 μ sec, but that of 1.09 μ was shorter than 100 n sec which is
the time constant of the measuring system.   As reported by
Gouterman[5] the unpaired d electron of the center atom coupled to
the π electron system gives rise to the splitting of triplet state
into tripdoublet and quartet.   Therefore the emission band at
1.12 μ may be due to the emission from quartet and the band at 1.09
μ to the emission from tripdoublet.
     The life time of the quartet was longer than that of tripdoublet
by several orders, indicating that the transition between quartet and
ground state is forbidden.   With increasing temperature, the
quartet emission decreased because of the increased nonradiative

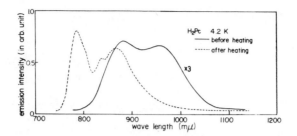

Fig.5   Emission spectra of H2Pc thin film
        before and after the heat treatment

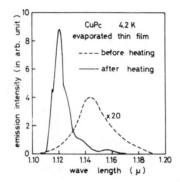

Fig.6   Emission spectra of CuPc thin film
        before and after the heat treatment

transition to the ground state.   On the other hand, the emission
from tripdoublet increased by raising temperature because of the
increased population of this state by thermal excitation from
quartet.   The emission spectra of evaporated thin films of
phthalocyanines were also studied.   As shown in Fig.5, the
original sample of H_2Pc thin film has a broad emission, being different markedly from the emission spectra of the single crystal
(Fig.2).   After heating the original sample to 280°C for
several hours in an evacuated quartz, the emission band shifted
to the shorter wave length side resulting in the similar spectra to
single crystals.   The change of emission spectra of CuPc by the
heat treatment are shown in Fig.6 also.
        Similar spectral changes by the heat treatment were also ob-
served in ZnPc and PtPc evaporated thin films. These results suggest

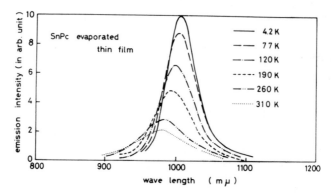

Fig.7   Temperature dependence of the emission
        spectra of SnPc thin film

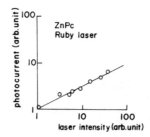

Fig.8   Dependence of induced current in ZnPc
        on ruby laser intensity

the phase transition from the α (original film) to the β (heat
treated film) phase by the heat treatment.

The temperature dependence of the emission spectra of the SnPc
thin film is shown in Fig.7.   With increasing temperature, the
emission peak shifted to the shorter wave length side and the emission intensity decreased.   This spectral change with temperature
resembles the one characteristic of excimer emission.
On the contrary to CuPc etc. the spectral change by heat treatment
was not observed in this SnPc film.   Therefore the crystal
structure of SnPc must be different from those of H_2Pc and the
phase transition should not occur by the heat treatment at 300°C.

Ruby and glass laser light induced photoconductivity in ZnPc
single crystal was also studied.   The relation between photocurrent and ruby laser light showed the square root dependence as

indicated in Fig.8.   In the case of glass laser (1.06 μ)
excitation, however, this dependence is nearly linear.   The ruby
laser line coincides with the singlet-singlet absorption edge of
phthalocyanine and the glass laser line lies nearly in the above
mentioned infrared photoresponse region due to singlet-triplet
absorption.

The carrier life time decreased with increasing the excitation
intensity, suggesting the bimolecular recombination process.
Therefore the singlet-singlet absorption as a carrier generation
mechanism and the bimolecular process as a carrier recombination
process may explain satisfactorily the sublinear dependence for ruby
laser excitation.   In the case of glass laser excitation, the
carrier generation due to the singlet-triplet absorption followed
by the triplet exciton-triplet exciton interaction plus the bimolecular recombination process should explain the linear dependence.

Sublinear dependence of photocurrent on light intensity was
also observed by several authers and explained by the exciton-trap
collision with exponential trap distribution[2], or the exciton-trap
collision with bimolecular recombination.[6]   In our case where
the high excitation level was attained by the intense laser beam,
the triplet exciton-triplet exciton collision seems to become
remarkable and exceed the triplet exciton-trap collision because of
the high density of triplet excitons.

REFERENCES

1) P. Day and R.J.P. Williams,   J. Chem. Phys. 42 4049 (1965)
2) S.E. Harrison,   J. Chem. Phys. 50 4739 (1969)
3) P.A. Barrett, C.E. Dent and R.P. Linstead,   J. Chem. Soc. 1936
   1719 (1936)
4) R.S. Becker and M. Kasha, J. Amer. Chem. Soc. 77 3659 (1955)
5) R.L. Ake and M. Gouterman,   Theoret. Chim. Acta. 15 20 (1969)
6) K. Gamo, K. Masuda and J. Yamaguchi,   J. Phys. Soc. Japan 25
   431 (1968)

OPTICAL AND ELECTRON PARAMAGNETIC PROPERTIES OF RADICALS

IN NAPHTHALENE AND ANTHRACENE CRYSTALS

Noriaki ITOH and Taisu CHONG

Department of Nuclear Engineering, Nagoya University

Furo-cho, Chikusa-ku, Nagoya, Japan

I. INTRODUCTION

The impurity and defect states in organic crystals have been studied by several authors[1],[2] particularly for the purpose of clarifying the energy transfer processes in organic materials. Most studies have been made for aromatic hydrocarbon crystals which include another aromatic hydrocarbon molecule as an impurity. In such cases the highest electronic state of the impurity is fully occupied. On the other hand the radical with an additional hydrogen atom to anthracene or naphthalene molecules (the cyclohexadienyl-type radical, referred as C-radical) and with a missing hydrogen atom from these molecules (the aryl-type radical, referred as A-radical) has a half-occupied orbital and the crystal containing these radicals would offer an interesting system to study the interaction between the localized state and the host lattice in organic crystals.

In the present paper the recent works on the radicals in naphthalene and anthracene crystals are reviewed. Optical absorption and luminescence caused by the radicals are identified and the interaction between the host lattice and the localized states is discussed.

II. IDENTIFICATION OF THE RADICAL

Electron paramagnetic resonance (EPR) has been the most powerful tool for the identification of the radical. The C-radicals have been identified clearly, since each hyperfine line was iden-

Table I. Radicals in Naphthalene and Anthracene Crystals

Host	Radicals		Thermal decay time constant at room temperature (hrs)
Naphthalene	C-radical	1-hydronaphthyl radical	very stable
		2-hydronaphthyl radical	4.0
	A-radical	naphthyl radical	11
Anthracene	C-radical	1-dibenzo-cyclo-hexadienyl radical	1.0
		9-dibenzo-cyclo-hexadienyl radical	very stable
	A-radical	anthracyl radical	unstable

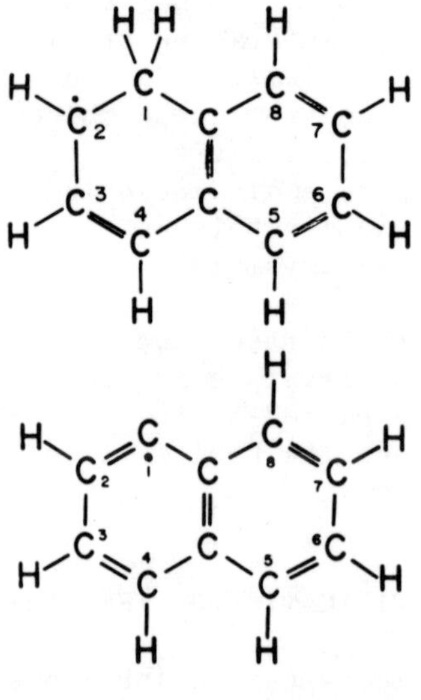

Fig. 1.   1-hydronaphthyl radical (upper) and 1-naphthyl radical (lower).

Table II. Optical Data of the Radicals in Naphthalene and
Anthracene at 80 K

Radical	Transition energy (eV)		Half width (eV)
	experiment	theory[a]	experiment
1-Hydronaph-thyl radical	2.31	2.40	0.011
	3.26	2.98	0.2
	3.68	3.68	0.3
		3.75	
2-Hydronaph-thyl radical	1.96	2.09	0.037
	3.06 [b]	3.13	–
	3.54 [b]	3.34	–
1-Dibenzo cyclohexa-dienyl radical	1.83	–	0.021
9-Dibenzo cyclohexa-dienyl radical	2.32	–	0.011

a) According to Hanazaki (reference 11)
b) Not uncertain

tified at some magnetic-field orientations and the angular dependence of each hyperfine line has been analyzed.[3]-[5] On the other
hand the A-radicals have broader EPR lines[6]-[8] than those observed in isolated systems[9] and their identification is not
straight forward. Circumstantial evidence, however, has been
presented to support that the primary radiation products are a
pair of the C-radical and the A-radical.[6],[7] Table I lists the
radicals of which the existence has been identified most clearly
and also their thermal stabilities at room temperature. Figure I
shows the structure of 1-hydronaphthyl and 1-naphthyl radicals as
examples of the C- and A- radicals, respectively.

III. ELECTRONIC STATES OF THE RADICAL

The lowest electronic transitions in the C-radicals show
sharp absorption bands accompanied with the **vibronic transitions.**[10]
Table II shows the experimental transition energies and the half
widths. It also includes the theoretical transition energies in

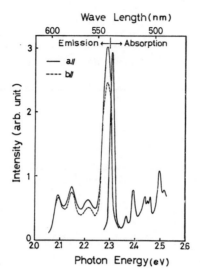

Fig. 2.   Emission and absorption spectra of the l-hydronaphthyl
          radical at liquid nitrogen temperature, excited at 337 nm.
          The solid and dashed curves show the a- and b- polarized
          emission, respectively.   The emission band is somewhat
          broader than the absorption band, since the resolution
          of the detection system was as large as 6 nm.

the hydronaphthyl radicals, calculated by Hanazaki[11] with the
Praiser-Parr-Pople method. Good agreement between experimental
and theoretical transition energies is obtained.

    The electronic states of the hydronaphthyl radical have been
described as the combination of the carbon π-orbitals and a
pseudo-π orbital formed by two hydrogens which lie in parallel
with the carbon π-orbitals. Since the ionization potential of
the pseudo-π orbital is smaller than that of the carbon-π orbital,[12] the energy state for an electron bound most loosely in the
radical should lie within the forbidden gap. The other electrons
would take the energy states which lie closely to those of valence
electrons. Therefore the C-radical is analogous to a donor state
in the semiconductors. On the other hand, the A-radical has a
missing electron in the σ-orbital which lies slightly above the

π−orbital.[13)] Thus the A−radical is considered to be a localized
positive hole center and is analogous to an accepter in semicon−
ductors. The effect of the radicals on the photoelectric and
space charge limited current are being studied.

The luminescence emitted by the excitation of the 1−hydro−
naphthyl radical has been observed by Shibata and Chong.[14)] As
shown in Fig. 2 the emission spectrum makes a mirror image of the
vibronic structures of the absorption band at 2.31 eV. The quan−
tum yield of the emission caused by the excitation at 3.68 eV
band is found to be the order of 10^{-3}. The life time of electrons
at the lowest excited state is about 80 nsec at liquid nitrogen
temperature and decreases slightly as the temperature is raised.
The analysis of the temperature dependence of the life time and
the luminescence intensity indicates that there is a thermally
activated quenching process with an activation energy of 0.06 eV.
The studies of the emission bands have not been extended to other
radicals but it is plausible that the C−radicals in anthracene
have similar emission bands.

IV. INTERACTION OF THE ELECTRONIC STATES OF THE RADICAL WITH
                    ENVIRONMENT

The wave functions of the electronic state of the C−radical
would have a weak coupling to the host molecule. The radical−
molecule resonance integral would be different from the resonance
integral[15)] between an excess electron in a host molecule and a
neighboring molecule by the following two reasons: (i) the pseudo−
π orbital gives additional resonance integrals: and (ii) there
are differences between the probability densities of the π−elect−
rons in each carbon π orbitals of the radical and those of the
molecule. The mixing of the wave functions for the electrons in
the radical through the resonance interaction between the pseudo−
π orbital and the carbon π orbital would be small owing to the
large difference in the ground state energies of the two orbitals.
Therefore the process (ii) would be predominant in producing the
mixing of the radical states with environment and the detailed
study of the electronic couping between the radical and the en−
vironment, such as an ENDOR study, would give informations on the
resonance integral between the carbon π−orbitals belonging neigh−
boring molecules.

Recently Nakagawa[16)] has found the phonon structures in the
2.31 eV absorption band to the lowest excited state of the 1−
hydronaphthyl radical and in each vibronic band associated with
it. This result indicates that the first excited state of the
radical has a weak coupling with the lattice. The structures of
the absorption band are in accordance with the phonon spectrum

calculated by Pawley[17] and the Raman spectrum [18] of the naphthalene crystal, indicating that the lattice vibrations are not distorted very much by the presence of the radical. The detailed
analysis of the spectrum is in progress.

The transitions to the higher excited states of the 1-hydro-
naphthyl radical have oscillator strengths larger than that of
2.31 eV band by more than an order of magnitude, and show broad
bands without structures. It follows that the lattice coupling
of the higher excited states is larger than that of the first
excited state. It is considered that the wave functions for the
higher excited states spread to the neighboring molecules by a
considerable degree.

V. INTERACTION BETWEEN RADICALS AND EXCITON

Effect of magnetic field on the recombination luminescence
has provided beautiful evidence of the interaction between the
triplet exciton and a radical.[19] The luminescence of the hydronaphthyl radical in naphthalene crystal is observed also by the
exciton excitation, the intensity being proportional to the radical concentration.[13] This result indicates that there is an
energy transfer from a host molecule to the radical. It was also
shown that the decay time of the singlet exciton luminescence decreases with increasing radical concentration. Since the configuration and electronic states of the C- and A-radicals are known
very well, the radical in the molecule would provide an interesting system to study the energy transfer processes in crystal.

The authors express their gratitude to T. Nakayama, Y. Shibata, K. Nakagawa and M. Higuchi for their helpful discussions.

REFERENCES

1) S. A. Rice and J. Jortner, Physics and Chemistry of Organic
   Solid State, vol. 8, ed. D. Fox, M. M. Labes, A. Weissberger
   (John Wiley, New York, 1967)p.199.
2) H. C. Wolf, Advances in Atomic and Molecular Physics, vol.1,
   ed. D. R. Bates and I. Esterman (Academic Press, New York,
   1967) p.119.
3) L. A. Harrah and R. C. Hughes, Molecular Crystals 5, 141 (1968).
4) T. Inoue, J. Phys. Soc. Japan 25, 914 (1968).
5) N. Itoh and T. Okubo, Molecular Crystals and Liquid Crystals
   17, 303 (1972).
6) Y. Akasaka, K. Masuda and S. Namba, J. Phys. Soc. Japan 30,
   1686 (1971).
7) T. Chong and N. Itoh, J. Phys. Soc. Japan 35, 518 (1973).

8) R. V. Lloyd, F. Magnotta and D. E. Wood, J. American Chem. Soc.
    90, 7142 (1968).
9) P.H. Kasai, E. Hedaya and E.B. Whipple, J. American Chem.
    Soc. 91, 4364 (1969).
10) T. Chong and N. Itoh, Molecular Crystals and Liquid Crystals
    11, 315 (1970).
11) T. Shida and I. Hanazaki, Bull. Chem. Soc. 43, 636 (1970).
12) I. Hanazaki and S. Nagakura, Bull. Chem. Soc. 38, 1298 (1965).
13) P.H. Ka ai, P.A. Clark and E.B. Whipple, J. American Chem.
    Soc. 92, 2640 (1970).
14) T. Chong, Y. Shibata and N. Itoh, to be published.
15) J.L. Katz, S.A. Rice, S. Choi and J. Jortner, J. Chem. Phys.
    39, 1683 (1963).
16) K. Nakagawa and N. Itoh, to be published.
17) G.S. Pawley, phys. stat. sol. 20, 347 (1967).
18) H. Moser and D. Stieler, Zeits f. Angewandte Physik 12, 280
    (1960).
19) P. Avakian and R.E. Merrifield, Molecular Crystals 5, 9 (1968).

CONTRIBUTION OF FREE RADICALS TO ELECTRICAL CONDUCTIVITY

Kohzoh Masuda and Susumu Namba

Faculty of Engineering Science, Osaka University

Toyonaka, Osaka

Electrical conductivity due to unpaired electrons is discussed here.

In order to get a clear understanding of the conduction mechanism due to the unpaired electrons (which means conduction due to the spins) testing an application of the hopping conduction mechanism to some particular cases is useful.

The hopping conduction dependence upon the exchange effect of the unpaired electron in free radicals has been investigated.[1] In this case, the dominant part of the conductivity is expressed to be

$$\sigma = n\frac{d^2 e^2}{kT}\, \nu \exp\left(-\frac{E}{kT}\right), \quad ---- \quad (1)$$

where n is the density of the unpaired electrons, d the mean distance between the spins, ν the trial time for hopping and E the activation energy for hopping.

Since the movement of the unpaired spin can be observed by the change of the line width of ESR the trial time of ν can be replaced by J/kT. J can be obtained from the exchange narrowing effect of ESR signal. The site to be filled by the hopped unpaired electron should be empty. This condition is assumed to be effectively satisfied by sufficiently strong interaction forces due to sorrounding atoms in the molecule.

The agreement between the calculated values and experimental values is shown in Fig. 1.

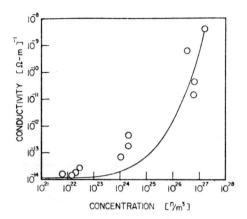

Fig. 1.   Concentration dependence of conductivity.
Circles indicate experimental values and full line
is of the calculated values.

The activation energy in equation (1) for the hopping is
supposed to be affected by sorrounded media. In the case of DPPH
crystal the solvent remaining in the crystal is one of such
materials.[2]

It is very difficult to change the amount of polarizability
without changing the structure of the molecule and crystal.

Various kinds of phthalocyanine crystals (H_2-Pc, Cu-Pc, Ni-Pc
and Zn-Pc) are used to study the effect of the polarization of the
metal ion located at the center of the molecule. Little differences
of structure are expected among these phthalocyanine crystals as
shown in Table I. The elementary analyses of the phthalocyanine
single crystals are also shown in Table II.

Table I.   The dimensions of unit
cell of phthalocyanine.

	a(Å)	b(Å)	c(Å)	β	space group
H_2-Pc	19.9	4.72	14.8	122°12′	$P_{21/a}$
Ni-Pc	19.9	4.71	14.9	121°54′	$P_{21/a}$
Cu-Pc	19.6	4.79	14.6	120°36′	$P_{21/a}$
Zn-Pc	19.22	4.87	14.52	120°02′	$P_{21/a}$

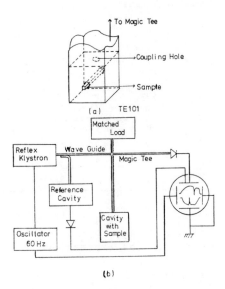

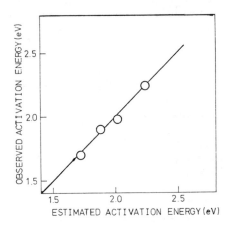

Fig. 2.   (a) The measurement cavity with a sample.
(b) Block diagram of the apparatus for the
measurement of dielectric constant (the frequency
is 9.5 GHz).

Fig. 3.   The estimated activation energy and
observed activation energy.

Table II. The elementary analysis of the
phthalocyanine single crystals.

		C(%)	H(%)	N(%)
$C_{32}N_8H_{18}$	Obs.	74.79	3.61	21.73
	Calc.	74.70	3.53	21.78
$C_{32}N_8H_{16}Zn$	Obs.	66.48	2.81	19.40
	Calc.	66.51	2.79	19.39
$C_{32}N_8H_{16}Cu$	Obs.	66.66	2.87	19.34
	Calc.	66.72	2.78	19.45
$C_{32}N_8H_{16}Ni$	Obs.	67.18	2.93	19.69
	Calc.	67.28	2.82	19.62

The activation energy in the expression of

$$\sigma = \sigma_0 \exp\left(-\frac{E}{kT}\right) \qquad ---- \qquad (2)$$

can be evaluated as

$$2E = I - 2P \qquad ---- \qquad (3)$$

where I is the ionization energy, and P the polarization energy.

In the various kind of phthalocyanine crystals, the difference
of activation energy of each phthalocyanine caould be explained
by using eq (3). The polarization energy could be calculated by
the following equation

$$P = \sum_n \frac{\alpha e^2}{2R_n^4} \qquad ---- \qquad (4)$$

where α is the polarizability, R_n the distance between the metal
ion and the molecule to be polarized. The polarizability is
calculated from the measured value of the dielectric constant by
means of the Clausius–Mosotti equation. The set up for the
dielectric constant measurement is shown in Fig. 2. The agreement
between observed values and estimated values of the activation
energies are shown in Fig. 3.[3]

In conclusion, there is great possibility of modulating the
conductivity due to unpaired electrons (spin) by locating the
atoms or molecules in some particular sites[4] in the crystal.

REFERENCES

1) K. Masuda and J. Yamaguchi: J. Phys. Soc. Japan $\underline{20}$ 1340 (1965).
2) K. Gamo, K. Masuda and J. Yamaguchi: J. Phys. Soc. of Japan $\underline{22}$ 1032 (1967).
3) Y. Aoyagi, K. Masuda and S. Namba: J. Phys. Soc. of Japan $\underline{31}$ 164 (1971).
4) W.A. Little, this volume, p. 145.

III. THEORY

GREEN FUNCTION OF AN EXCITON COUPLED WITH PHONONS IN MOLECULAR CRYSTALS

Kaoru Iguchi

Department of Chemistry, Waseda University

4-170, Nishiokubo, Shinjuku, Tokyo, Japan

The Hamiltonian of an exciton interacting with phonons in molecular crystals is assumed as

$$H = \varepsilon \sum_n a_n^+ a_n + \sum_{mn}{}' J_{mn} a_m^+ a_n + \sum_q \hbar \omega_q b_q^+ b_q$$
$$+ N^{-1/2} \sum_q \sum_n \left\{ K_{nq} a_n^+ a_n + \sum_m{}' L_{mnq} a_m^+ a_n \right\} (b_{-q}^+ + b_q),$$

(1)

where a_n^+, a_n are exciton creation and annihilation operators at site n, and b_q^+, b_q are those of phonon with a wave vector q. Coefficients J_{mn}, K_{nq} and L_{mnq} are so-called resonance, exciton-vibration and hopping interactions. We take a wave-representation by the transformation :

$$a_n = N^{-1/2} \sum_K a_K \exp(iKnd), \quad a_n^+ = N^{-1/2} \sum_K a_K^+ \exp(-iKnd)$$

with the assumption

$$K_{nq} = K_q \exp(iqnd), \quad L_{mnq} = L_q(|m-n|) \exp\{iq(m-n)d\},$$

and we have

$$H = H_0 + H',$$
$$H_0 = \sum_K \varepsilon_K a_K^+ a_K + \sum_q \hbar \omega_q b_q^+ b_q,$$
$$H' = N^{-1/2} \sum_K \sum_q C(q) a_{K+q}^+ a_K (b_{-q}^+ + b_q),$$

(2)

where

$$\varepsilon_K = \varepsilon + 2 \sum_{s>0} J_s \cos(Skd), \quad C(q) = K_q + 2 \sum_{s>0} L_q(s),$$

and d is the lattice vector.

The Green function of an exciton at a time t after the creation at time zero is given by[1]

$$G_k(t) = -i\,T\langle 0; \cdots \nu_q \cdots | a_k(t) U(t,0) a_k^+ | 0; \cdots \nu_q \cdots \rangle \tag{3}$$

where

$$U(t,0) = T \exp\left[-\frac{i}{\hbar} \int_0^t \exp(iH_0 t'/\hbar) H' \exp(-iH_0 t'/\hbar)\,dt'\right] \tag{4}$$

and $|0; \cdots \nu_q \cdots \rangle$ means a zero-exciton state with a quantum number ν_q of phonon for wave-vector q.

Putting

$$\Psi = a_k^+ |0; \cdots \nu_q \cdots \rangle, \quad V(t) = \exp(iH_0 t/\hbar) H' \exp(-iH_0 t/\hbar),$$

we have

$$\langle \Psi | U(t,0) \Psi \rangle = \sum_{n=0}^{\infty} \left(-\frac{i}{\hbar}\right)^n \frac{1}{n!} \int_0^t \cdots \int_0^t T \langle \Psi | V(t_1) V(t_2) \cdots V(t_n) | \Psi \rangle\, dt_1 \cdots dt_n$$

$$= \sum_{n=0}^{\infty} \left(-\frac{i}{\hbar}\right)^n \frac{1}{n!} \langle [w]^n \rangle \tag{5}$$

in symbolic notation.   After a calculation we have

$$\langle [w] \rangle = 0,$$

$$\langle [w]^2 \rangle / 2! = \langle F \rangle = N^{-1} \sum_q C_{(q)}^2 \hbar \left\{ i A_{kq} t + B_{kq}^{(1)} \{ 1 - \exp(i \Omega_{kq}^- t/\hbar) \} \right.$$

$$\left. + B_{kq}^{(2)} \{ 1 - \exp(-i \Omega_{kq}^+ t/\hbar) \} \right\}, \tag{6}$$

where

$$\left.\begin{array}{l}
A_{kq} = \bar{\nu}_q / \hbar \Omega_{kq}^- - (\bar{\nu}_q + 1)/\hbar \Omega_{kq}^+, \quad \bar{\nu}_q = \{ \exp(\hbar \omega_q / kT) - 1 \}^{-1} \\[4pt]
B_{kq}^{(1)} = \bar{\nu}_q / \hbar (\Omega_{kq}^-)^2, \quad B_{kq}^{(2)} = (\bar{\nu}_q + 1)/\hbar (\Omega_{kq}^+)^2, \\[4pt]
\hbar \Omega_{kq}^+ = \hbar \omega_q + \varepsilon_{k+q} - \varepsilon_k, \quad \hbar \Omega_{kq}^- = \hbar \omega_q - \varepsilon_{k+q} + \varepsilon_k.
\end{array}\right\} \tag{7}$$

The $\langle [w]^4 \rangle$ term can be dissolved into contractions as

$$\langle [w]^4 \rangle = \int_0^t \cdots \int_0^t dt_1 \cdots dt_4\, T \{ \langle V(t_1) V(t_2) \rangle \langle V(t_3) V(t_4) \rangle +$$

$$\langle V(t_1) V(t_3) \rangle \langle V(t_2) V(t_4) \rangle + \langle V(t_1) V(t_4) \rangle \langle V(t_2) V(t_3) \rangle \}. \tag{8}$$

In the expansion of the right hand of eq.(8) we take up only those terms which can be expressed as the powers of $\langle F \rangle$.   This approximation is essentially the same as those employed by Toyozawa[2] or

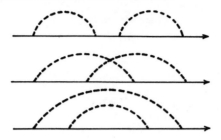

Fig.1 Feynman diagram of contractions in eq.(8) .

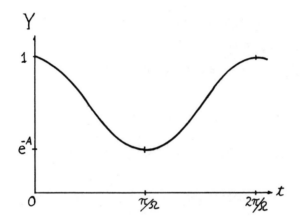

Fig.2 Function $Y = \exp\left\{-A\sin^2(\Omega t/2)\right\}$.

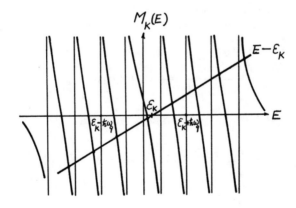

Fig.3 Graphical solution of $E - \mathcal{E}_k - M_k(E) = 0$.

Grover and Silbey[3]. Then we have

$$\langle [w]^4 \rangle = 3 \times 2^2 \times \langle F \rangle^2 ,$$

and generally

$$\langle [w]^{2n+1} \rangle = 0, \quad \langle [w]^{2n} \rangle / (2n)! = \langle F \rangle^n / n! ,$$

then

$$\langle \Psi | U(t,0) | \Psi \rangle = \sum_{n=0}^{\infty} (-\hbar^{-2})^n \langle F \rangle^n / n! = \exp(-\langle F \rangle \hbar^{-2}) . \tag{9}$$

Using $a_k(t) = a_k \exp(-i\varepsilon_k t/\hbar)$ we have finally

$$t < 0 \ \cdots \ G_K(t) = 0,$$

$$t > 0 \ \cdots \ G_K(t) = -i \exp[-i\varepsilon_k t/\hbar - \langle F \rangle / \hbar^2] \tag{10}$$

or, in more detail, for $t > 0$

$$G_K(t) = -i \exp(-i E_k t/\hbar) \times$$

$$\exp\left[\sum_g D_{1g} \bar{v}_g \{\exp(i\Omega^-_{kg} t) - 1\} + D_{2g}(\bar{v}_g + 1)\{\exp(-i\Omega^+_{kg} t) - 1\} \right] \tag{11}$$

where

$$E_k = \varepsilon_k + N^{-1} \sum_g C^2(g) \{ \bar{v}_g / \hbar \Omega^-_{kg} - (\bar{v}_g + 1) / \hbar \Omega^+_{kg} \} ,$$

$$D_{1g} = C^2(g) / N(\hbar \Omega^-_{kg})^2, \quad D_{2g} = C^2(g) / N(\hbar \Omega^+_{kg})^2 . \tag{12}$$

Expanding the complicated exponential function in the right-hand side of eq.(11) and reassembling we have

$$i G_K(t) \exp(i E_k t/\hbar) \eta_K^{-1}$$

$$= \prod_g \left(\sum_{r=0}^{\infty} \sum_{s=0}^{\infty} \frac{D_{1g}^r D_{2g}^s \bar{v}_g^r (\bar{v}_g + 1)^s}{r! \ s!} \exp\{i(r\Omega^-_{kg} - s\Omega^+_{kg})t\} \right) \tag{13}$$

$$= \prod_g \sum_{p=-\infty}^{\infty} (D_{2g}/D_{1g})^{p/2} \exp(p \beta \hbar \omega_g / 2 - i p \omega_g t) \times$$

$$I_p \left(2\{D_{1g} D_{2g} \bar{v}_g (\bar{v}_g + 1)\}^{1/2} \right) \exp[i(\varepsilon_k - \varepsilon_{k+g})t]) \tag{14}$$

where $I_p(x)$ is the modified Bessel function of order p and argument x, and $\beta = 1/kT$. Also

$$\eta_K = \exp\left[-\sum_g \{ D_{1g} \bar{v}_g + D_{2g}(\bar{v}_g + 1) \} \right]$$

Eq.(14) is an extension of the results of Gosar and Choi[4] or of Grover and Silbey[3], and coincides with them when $\varepsilon_k = \varepsilon_{k+g}$, i.e., the resonance interaction is neglected.

We rewrite eq.(13) as

$$i\,G_k(t) = \exp(-i\,E_k\,t/\hbar)\prod_q G_{kq}(t),$$

$$G_{kq}(t) = \exp\left[-2D_{1q}\bar{v}_q \sin^2(\Omega^-_{kq}\,t/2) - 2D_{2q}(\bar{v}_q+1)\sin^2(\Omega^+_{kq}\,t/2)\right]$$

(15)

The function $\exp\left\{-A\sin^2(\Omega t/2)\right\}$ in eq.(15) is a periodic function of t with a period $2\pi/\Omega$, and $G_{kq}(t)$ is a product of two such functions with different frequencies Ω^+_{kq} and Ω^-_{kq}. The oscillation of $G_{kq}(t)$ means the to-and fro interchange of energy and momentum between the exciton and phonon field.

The energy-argument expression of Green function is given by:

$$i\,G_k(E) = \frac{i}{2\pi}\int_{-\infty}^{\infty} G_k(t)\exp(i\,E\,t/\hbar)dt$$

$$= \frac{\eta_k}{2\pi}\int_0^{\infty} dt\cdot\exp\left\{i\frac{E-E_k}{\hbar}t\right\}\prod_q\sum_\ell R_{\ell q}\exp\left\{it(r\Omega^-_{kq} - s\Omega^+_{kq})\right\}$$

(16)

where ℓ means a pair of r and s, and $R_{\ell q} = \dfrac{D^r_{1q}\,D^s_{2q}\,\bar{v}_q(\bar{v}_q+1)^s}{r!\,s!}$.

Then we have

$$G_k(E+i0) = \frac{\eta_k\hbar}{2\pi}\sum_{\ell\text{-set}} I(\ell\text{-set})\left[E - E_k - \hbar\sum_j(S\Omega^+_{kq} - r\Omega^-_{kq})_j + i0\right]^{-1}$$

(17)

where ℓ-set means a set of ℓ_j's, and j numbers the wave-vectors of phonon, and $I(\ell\text{-set}) = \prod_q R_{\ell q}$ for an ℓ-set. The poles of $G_k(E)$ give the possible energies of exciton after a long elapse of time, For each ℓ-set the energy is given as:

$$E = E_k + \hbar\sum_j(S\Omega^+_{kq} - r\Omega^-_{kq})_j$$

$$= \varepsilon_k + N^{-1}\sum_j(C^2(q)A_{kq})_j + \sum_j\left\{(S-r)\hbar\omega_q - (S+r)(\varepsilon_k - \varepsilon_{k+q})\right\}_j.$$

(18)

The quantity $\hbar\Omega^+_{kq}$ is the increase of energy by virtual phonon-creation process; $\varepsilon_k \to \varepsilon_{k+q} + \hbar\omega_q$ and $\hbar\Omega^-_{kq}$ is the decrement by the phonon-annihilation process; $\varepsilon_k + \hbar\omega_q \to \varepsilon_{k+q}$. The resultant increase of phonon energy is $(s-r)\hbar\omega_q$, and $(s+r)(\varepsilon_k - \varepsilon_{k+q})$ is the resultant decrease of free-exciton energy, and $s+r$ is the number of relevant phonons. The term $C^2(q)A_{kq}$ is the energy shift due to the exciton-phonon interaction. Thus our method includes the multiple scatter - ing processes until infinity, though not perfectly. Also we see the effect of hopping term is only the replacement of K_q by $C(q)$.

We shall compare our result with those by other workers. By the equation-of-motion method of Zubarev[5] with the first-order decoupling approximation, the Green function with energy argument is obtained as

$$G_k(E) = (2\pi)^{-1}\left[E - \varepsilon_k - M_k(E)\right]^{-1}$$

(19)

and the causal Green function as :

$$G_{Tk}^+(t) = \frac{1}{2\pi} \int_{-\infty}^{\infty} \exp(-iEt/\hbar)\left[E - \mathcal{E}_k - M_k(E^+)\right]^{-1} dE$$

(20)

where $E^+ = E + 0$, and the self energy M_k is given as :

$$M_k(E^+) = M_k(E) - i\gamma_k(E) = P\sum_{\bar{g}} \frac{C^2(\bar{g})}{N}\left\{\frac{\bar{v}_{\bar{g}} + 1}{E - \mathcal{E}_{k-\bar{g}} - \hbar\omega_{\bar{g}}} + \frac{\bar{v}_{\bar{g}}}{E - \mathcal{E}_{k-\bar{g}} + \hbar\omega_{\bar{g}}}\right\}$$

$$- i\pi\sum_{\bar{g}} \frac{C^2(\bar{g})}{N}\left\{(\bar{v}_{\bar{g}} + 1)\delta(E - \mathcal{E}_{k-\bar{g}} - \hbar\omega_{\bar{g}}) + \bar{v}_{\bar{g}}\,\delta(E - \mathcal{E}_{k-\bar{g}} + \hbar\omega_{\bar{g}})\right\}.$$

(21)

The pole of $G_k(E)$ is given approximately by:

$$E_a \simeq \mathcal{E}_k + M_k(\mathcal{E}_k) - i\gamma_k(\mathcal{E}_k).$$

(22)

Then we have, for $t > 0$,

$$G_k^+(t) = -i\exp(-iE_a t/\hbar)$$

$$\simeq -i\exp\left[-i\{\mathcal{E}_k + M_k(\mathcal{E}_k)\}t/\hbar - \gamma_k(\mathcal{E}_k)t/\hbar\right].$$

(23)

We see that $\mathcal{E}_k + M_k(\mathcal{E}_k)$ is equal to E_k in eq. (11), and we may say the equation-of-motion method is of lower order in approximation than our perturbation expansiom method. Also the factor $\exp(-\gamma t/\hbar)$ means the exponential decay of probability to find the exciton in the k-state though our eq. (15) indicates a complicated oscillatory behavior.

By the calculation of Takeno[6] the pole of $G_k(E)$ is obtained as the root of the following equation :

$$E - \mathcal{E}_k = M_k(E) = \sum_{\bar{g}} \frac{C^2(\bar{g})}{N}\left\{\frac{\bar{v}_{\bar{g}} + 1}{E - \mathcal{E}_k - \hbar\omega_{\bar{g}}} + \frac{\bar{v}_{\bar{g}}}{E - \mathcal{E}_k + \hbar\omega_{\bar{g}}}\right\}.$$

(24)

The approximate zero-phonon and one-phonon energies are:

$$E_0 = \mathcal{E}_k - \sum_j C^2(\bar{g}_j)/N\hbar\omega_{\bar{g}_j},$$

$$E_{\bar{g}+} = \mathcal{E}_k + \hbar\omega_{\bar{g}} + C^2(\bar{g})(\bar{v}_{\bar{g}} + 1)/N\hbar\omega_{\bar{g}},$$

$$E_{\bar{g}-} = \mathcal{E}_k - \hbar\omega_{\bar{g}} - C^2(\bar{g})\bar{v}_{\bar{g}}/N\hbar\omega_{\bar{g}}.$$

(25)

Similar values can be obtained very simply from our eq. (18), putting

1. All $s = r = 0$, then $E = E_k = E_0$.
2. $s_1 = 1$, others $= 0$, then

$$E = E_k + \hbar\Omega_{k\bar{g}_1}^+ \simeq \mathcal{E}_{k-\bar{g}_1} + \hbar\omega_{\bar{g}_1} - \sum_j C^2(\bar{g}_j)/N\hbar\omega_{\bar{g}_j}$$

which corresponds to E_q^+.
3. $r_1 = 1$, others $= 0$, then

$$E = E_k - \hbar \Omega_{k g_1}^- \simeq \mathcal{E}_{k-g_1} - \hbar \omega_{g_1} - \sum_j C^2(g_j)) / N \hbar \omega_{g_j}$$

which corresponds to E_q^- .

REFERENCES

1. N.H.March, W.H.Young, S.Sampanthar : The Many-Body Problem in Quantum Mechanics (Cambridge University Press, 1967)
2. Y.Toyozawa, Prog.Theor.Phys. 20, 53(1958)
3. M.Grover and R.Silbey, J.Chem.Phys. 54, 4843(1971)
4. P.Gosar and S.Choi, Excitons, Magnons and Phonons in Molecular Crystals, edited by A.B.Zahlan(Cambridge University Press,1968)
5. Z.N.Zubarev, Soviet Phys.Uspekhi, 3, 320, (1960)
6. S.Takeno, J.Chem.Phys. 46, 2481(1967)

THEORY OF FRENKEL EXCITONS USING A TWO-LEVEL-ATOM MODEL

Shozo Takeno

Department of Nuclear Engineering, Kyoto University

Kyoto, Japan

The principal purpose of the present paper is to develop a formal, rigorous theory of Frenkel excitons with or without interaction with an external electro-magnetic field within the framework of a two-level-atom model. Employed implicitly but extensively in exciton problems have been model systems composed of interacting two-level atoms in which only two of energy levels of a given atom, corresponding to the ground state and an excited state, endowed with a transition dipole moment, are taken into account. The most common method traditionally employed here is to treat Frenkel as well as Wannier excitons simply as Bosons. It is however to be reminded that excitons, both Frenkel and Wannier types, are neither Bosons nor Fermions. Specifically, Frenkel excitons in such a model system are to be considered as Paulions or being equivalent to S=1/2 spins. Except for the specific case of one-dimensional system with nearest-neighbour interactions only,[1] the conventional procedure, in analogy with magnons in Heisenberg magnets, would be to regard excitons as a non-ideal Bose gas in which the deviation from the Bose statistics is expressed in the form of effective interactions among excitons, called kinematical interactions.[2],[3] The discussion along this line however is generally much involved. Furthermore, there exist intrinsic or dynamical exciton-exciton interactions, even if they are regarded as Paulions or, more generally, as electron-hole pairs. It is shown in this paper that within the framework of a two-level-atom model the model exciton Hamiltonian is rigorously expressed in the form of the Heisenberg model Hamiltonian for S=1/2 spins with long-range anisotropic exchange interactions, and that Frenkel excitons thus obtained are generally expressed as quantum-mechanical nonlinear polarization waves, the nonlinearity being characterized by the atomic level population.

To begin with, we consider a system of identical atoms. Let ε_f and φ_{if} be the natural energy eigenvalue and the corresponding eigenfunction of an atom at a site i specified by quantum number f. Also, let a_{if}^{+} and a_{if} be creation and annihilation operators of an electron in the state φ_{if}; Any two a's belonging to different atomic sites are taken to be commutable with each other. We take the Hamiltonain of the system to be of the form

$$H = \sum_{if} \varepsilon_f\, a_{if}^{+}\, a_{if} + \tfrac{1}{2} \sum_{ij} \sum_{ff'f''f'''} \langle ff'|V(ij)|f''f'''\rangle\, a_{if}^{+}\, a_{jf'}^{+}\, a_{jf''} a_{if'''} + \sum_{iff'} \langle f|V(i)|f'\rangle a_{if}^{+}\, a_{if'}$$

$$(1)$$

Here, $V(ij)$ and $V(i)$ denote the Coulomb interaction between the i and j atoms and the interaction of the i atom with an external field, respectively. The definition of the matrix elements are
$\langle ff'|V(ij)|f''f'''\rangle = \iint \varphi_{if}^{*}(r_i)\varphi_{jf'}^{*}(r_j) V(ij) \varphi_{if''}(r_i) \varphi_{jf'''}(r_j)\, dr_i\, dr_j$ and
$\langle f|V(i)|f'\rangle = \int \varphi_{if}^{*}(r_i) V(r_i)\varphi_{if'}(r_i) dr_i$. In what follows we take into account only two of energy states of a given atom, the ground state and an excited state under consideration, which are denoted by f=0 and f=1, respectively. We introduce exciton operators A_i^{+} and A_i by the equations

$$A_i^{+} = a_{i1}^{+} a_{io} = \sigma_i^{+} \qquad \text{and} \qquad A_i = a_{io}^{+} a_{i1} = \sigma_i^{-}, \qquad (2)$$

where $\vec{\sigma}_i = (\sigma_i^{x}, \sigma_i^{y}, \sigma_i^{z})$ with $\sigma_i^{\pm} = \sigma_i^{x} \pm \sigma_i^{y}$ are the Pauli operators. Several relations useful for later calculations are listed below:

$$A_i^{+} A_i = \sigma_i^{+}\sigma_i^{-} = \sigma_i^{z} + (\tfrac{1}{2}) \equiv n_{i1}, \qquad (3)$$

$$A_i A_i^{+} = \sigma_i^{-}\sigma_i^{+} = (\tfrac{1}{2}) - \sigma_i^{z} \equiv n_{io}, \qquad (4)$$

$$n_{io} + n_{i1} = A_i A_i^{+} + A_i^{+} A_i = \sigma_i^{-}\sigma_i^{+} + \sigma_i^{+}\sigma_i^{-} = 1 \qquad (5)$$

$$A_i A_j - A_j A_i = \sigma_i^{-}\sigma_j^{+} - \sigma_j^{+}\sigma_i^{-} = \Delta(ij)(1 - 2A_i^{+}A_i), \qquad (6)$$

where $\Delta(ij)$ is Kronecker's delta. Equation (1) with f=0 and 1 only, inconjuction with Eqs.(2)-(6), reduce to

$$H = \sum_{i} (\varepsilon_1 - \varepsilon_0 + P_i)\sigma_i^{z} + \tfrac{1}{2}\sum_{ij} J(ij)\{\sigma_i^{+}\sigma_j^{-} + \sigma_j^{+}\sigma_i^{-} + \sigma_i^{+}\sigma_j^{+} + \sigma_i^{-}\sigma_j^{-}\}$$

$$+ \tfrac{1}{2}\sum_{ij} I(ij)\sigma_i^{+}\sigma_j^{+}\sigma_i^{-}\sigma_j^{-} + \sum_{i}(v_i \sigma_i^{+} + v_i^{*}\sigma_i^{-}) \qquad (7)$$

or

$$H = \sum_{i} \omega_i\, \sigma_i^{z} + 2\sum_{ij} J(ij)\, \sigma_i^{x}\sigma_j^{x} + \tfrac{1}{2}\sum_{ij} I(ij)\,\sigma_i^{z}\sigma_j^{z} + 2\sum_{i} v_i\, \sigma_{ij}^{x}, \qquad (8)$$

where

$$\omega_i = \mathcal{E}_1 - \mathcal{E}_0 + D_i + (\tfrac{1}{2}) \sum_j I(ij), \quad D_i = \sum_j \{ \langle 10 | \mathcal{V}(ij) | 10 \rangle - \langle 00 | \mathcal{V}(ij) | 00 \rangle \}$$

$$\mathcal{J}(ij) = \langle 10 | \mathcal{V}(ij) | 01 \rangle,$$

$$I(ij) = \langle 11 | \mathcal{V}(ij) | 11 \rangle - 2 \langle 01 | \mathcal{V}(ij) | 01 \rangle + \langle 00 | \mathcal{V}(ij) | 00 \rangle,$$

$$\mathcal{V}_i = \mathcal{V}_i^* = \langle 1 | \mathcal{V}(i) | 0 \rangle = \langle 0 | \mathcal{V}(i) | 1 \rangle \qquad (9)$$

In arriving at the above results, it has been assumed that the exchange interaction of electrons belonging to different atoms are negligible.

The physical meaning of the result obtained above can be understood more clearly and intuitively by using the dipole approximation in describing the exciton-field interaction, treating the external field as a c-number:

$$v_i = -\mu \vec{e}_i \cdot \vec{E}_i \qquad \text{or} \quad v_i (\sigma_i^+ + \sigma_i^-) = -p_i \vec{e}_i \cdot \vec{E}_i \qquad \text{with } p_i = 2\mu \sigma_i^x, \qquad (10)$$

and by observing the realtion

$$\sigma_i^z = (n_{i1} - n_{i0}) \equiv n_i / 2. \qquad (11)$$

In Eq. (10) μ is the matrix element of the atomic dipole moment $\vec{\mu}_i$ of the i atom, $\vec{e}_i$ is a unit vector in the direction of $\vec{\mu}_i$ and $\vec{E}_i$ is an external field seen by the i atom. The quantity

$$\vec{p}_i \equiv p_i \vec{e}_i = \int \Psi_*(r_i) \, \vec{\mu}_i \, \Psi(r_i) dr_i \qquad (12)$$

is the expectation value of $\vec{\mu}_i$ with respect to the quantized atomic wave function

$$\Psi(r_i) = a_{i0} \varphi_{i0}(r_i) + a_{i1} \varphi_{i1}(r_i). \qquad (13)$$

Thus, the quantities σ_i^x and σ_i^z are intimately connected with the atomic dipole moment operator P_i and the population difference n_i defined by Eq. (11).

From Eq. (8) equations of motion for the σ's can be obtained in the form of the Bloch equation

$$d\vec{\sigma}_i / dt = \vec{\Omega}_i \times \vec{\sigma}_i \qquad \text{with} \quad \vec{\Omega}_i = (2v_i^*, 0, \omega_i^*), \qquad (14)$$

where

$$v_i^* = v_i + 2\Sigma_j J(ij)\sigma_j^x = -\mu\mathcal{E}_i^*, \qquad \omega_i^* = \omega_i + \sum_j I(ij)\sigma_j^z,$$

$$\mathcal{E}_i^* = \mathcal{E}_i - \sum_j T(ij)P_j, \quad \text{with } T(ij) = J(ij)/\mu^2 \text{ and } \mathcal{E}_i = \vec{e}_i \cdot \vec{E}_i. \tag{15}$$

Equation (14) can be considered as a generalization of the result first obtained by Feynman and Vernon in the case of non-interacting two-level atoms.[4] Equations of motion obeyed by P_i and n_i are obtained in a straightforward manner from Eq. (14) as follows

$$\frac{d^2 P_i}{dt^2} + \gamma(\omega_i^*)\frac{dP_i}{dt} + \omega_i^{*2}P_i - 2\mu^2\omega_i^* n_i \sum_j T(ij)P_j = -2\mu^2\omega_i^* n_i \mathcal{E}_i \tag{16}$$

$$\frac{\omega_i^*}{2}\frac{dn_i}{dt} = \frac{dP_i}{dt}\mathcal{E}_i^*, \tag{17}$$

where $\gamma(\omega_i^*) = -(d\omega_i^*/dt)(\omega_i^*)^{-1}$. These equations are exactly equivalent to Eq. (14), but in the form (16) and (17) we have a simple physical interpretation. The atomic dipole moment responds to an applied field $\mathcal{E}_i$ according to a driven polarization-wave equation, with the feature that the coupling constant $-2\mu^2\omega_i^* n_i$ and the term $-2\mu^2\omega_i^* n_i \sum_j T(ij)p_j$ are proportional to the population difference n_i, reversing sign when n_i passes through zero. Equation (17) is simply a statement of conservation of energy. There is one further immediate consequence of writing the equations in this form: Multiplying Eq. (16) by dp_i/dt from the right, substituting Eq. (17) and neglecting the time variation of ω_i^* and the non-commutativity of p_i and dp_i/dt, we get after integrating:

$$\left(\frac{dP_i}{dt}\right)^2 + \omega_i^{*2}P_i^2 + \omega_i^{*2}\mu^2 n_i^2 = \mu^2\omega_i^{*2}, \tag{18}$$

where we have determined the constant of integration by requiring that $(dp_i/dt)_{t=0} = (p_i)_{t=0} = 0$ and $(n_i)_{t=0} = \pm 1$. Equation (16) can be recast into the form

$$\frac{d^2 P_i}{dt^2} + \gamma(\omega_i^*)\frac{dP_i}{dt} + \omega_i^{*2}P_i + \alpha_i(0)\omega_i^{*2}(-n_i)\sum_j T(ij)P_j = \alpha_i(0)\omega_i^{*2}(-n_i)\mathcal{E}_i, \tag{19}$$

where

$$\alpha_i(\omega) = 2\mu^2\omega_i^*/(\omega_i^{*2} - \omega^2) \tag{20}$$

is the polarizability of a two-level atom in isolation with the energy separation ω_i^*. Equation (16) or (19) and (17) generally describe "nonlinear" Frenkel excitons as quantum-mechanical nonlinear polarization waves, the nonlinearity being characterized by the atomic level population. It is of interest to note that Eq.(19) is exactly identical with an equation obtainable from the classical Drude-Lorentz model, provided the factor $-n_i$ is replaced by unity. The energy eigenvalues $\omega(k)$ as a function of wave vector k are obtained from Eq.(16), assuming the spatial arrangement of atoms in the system to be periodic and by putting $P_i = p e^{i(\omega t - k \cdot i)}$ as follows:

$$\omega \equiv \omega(k) = \left[\omega_0^{*2} + 2\mu^2 \omega_0^*(-n) T(k) + i\omega \gamma(\omega_0^*) \right]^{1/2} . \qquad (21)$$

Here, T(k) is the Fourier transform of the dipole-dipole interaction energy T(ij) or dipole sums, and we have neglected the i-dependence of ω_i^* and n_i, rewriting these as ω_0^* and n, respectively. While, the corresponding eigenvalues derivable from the naive form of the model exciton Hamiltonian based on the simplified Heitler-London model is given by $\omega(k) = \epsilon_1 - \epsilon_0 + \mu^2 T(k)$. The above results differ from those obtained in conventional linear exciton theories in two respects. The one is the appearance of the population difference n, which introduces into the problem a nonlinearity, being of fundamental importance in our present case as well as in the case of nonlinear optics and laser or maser physics in which systems composed of non-interacting two-level atoms have been employed extensively as a working model. The other is the inclusion of the dynamical exciton-exciton interaction characterized by the quantity I(ij), which gives rise to the shift or the modulation of the excitation energy of an atom.

Aside from the Frenkel-exciton problem in general, the results obtained above can be applied to several problems. Exciton-photon coupled modes can be obtained in a straightforward manner by treating Eqs.(16) and (17) simultaneously with the Maxwell equation. It is shown that conventional polariton modes appear only in certain limiting cases of solutions of these coupled equations. It is also of interest to note that Eq.(20) with $\gamma(\omega_i^*)$ and $-n_i$ replaced by zero and unity, respectively, has a wide variety of applicability for the calculation of the van der Waals forces. Underlying fact here is that it can be recognized as the decrease in the sum of the zero-point energies of Frenkel excitons upon formation of atomic or molecular aggregate.

REFERENCES

1) D.B. Chesnut and A. Suna, J. Chem. Phys. 39, 146 (1963).
2) F.J. Dyson, Phys. Rev. 102, 1217 (1956).
3) V.M. Agranovich and B.S. Toshich, Soviet Phys.-JETP 26, 104 (1968).
4) R.P. Feynman and F.L. Vernon, J. Appl. Phys. 28, 49 (1957).

IV. CRYSTALLIZATION

GLASS-FORMING PHOTOCONDUCTIVE ORGANIC COMPOUNDS. I.
PHASE CHANGE AND SINGLE CRYSTAL GROWTH OF 1,3-DIPHENYL-5-
(p-CHLOROPHENYL)-PYRAZOLINE

K. Kato, M. Yokoyama, Y. Shirota, H. Mikawa,
M. Sorai*, H. Suga*, and S. Seki*
Department of Applied Chemistry, Faculty of Engineering,
Osaka University, Yamadakami, Suita, Osaka, Japan
* Department of Chemistry, Faculty of Science,
Toyonaka, Osaka, Japan

It is well known that linear polymers regarded as amorphous
solids generally keep a stable glassy state for quite a long time
at around room temperature or above. Some low-molecular-weight
organic compounds have also been reported to form stable glasses
at around room temperature or above, where either the three-
dimensional hydrogen-bonding force or only van der Waals force is
operative. A branched, propeller-like hydrocarbon, e.g., 1,3,5-
tri-α-naphthylbenzene, is a typical example of a van der Waals
glass former, which is reported to form a glass spontaneously on
cooling with a glass transition temperature at about 70°C (1).

We have found that 1,3-diphenyl-5-(p-chlorophenyl)-pyrazoline
(I), which shows photoconductivity (2), supercools very readily
and forms a very stable glass at around room temperature. Indeed,
the glassy state of this substance has been stable for years in
our laboratory.

(I)

77

There has currently been an increasing interest in the
investigation of the effect of structure, i.e., crystalline,
amorphous or liquid state, on the charge carrier transport (3).
If a large single crystal of the pyrazoline derivative I is
successfully grown, it will allow direct comparison of the optical
and photoconductive properties of this material between in the
crystalline and the glassy state at the same temperature range.
This paper describes the phase change, and the successful zone
refining and single crystal growth of I. This seems to be the
first example of the growth of a large single crystal from the
melt with respect to the glass-forming system.

PHASE CHANGE

Figure 1 shows the results of the differential thermal
analysis (DTA) of the crystalline and glassy state samples of I.
As is evident from the Figure, the glass transformation was
observed at around $16.5 - 21.5°C$ with regard to the glassy state
sample, and above that temperature region this substance keeps a
supercooled liquid state. Somewhat exothermic behavior observed
at above 80°C may be due to the partial crystallization and at
ca. 130°C a few endothermic peaks were obsrved due to the melting.
The results of dilatometric and penetrometric measurements are in
good agreement with those of DTA. The dilatometric measurement
of the volume-temperature dependence indicated that the supercooled
liquid changes into the glassy solid below ca. 28°C, and the
penetrometric test showed that the supercooled liquid becomes very
hard below 30°C. These results are shown in Figure 2. The
behavior of the phase change of I is depicted in Scheme 1.

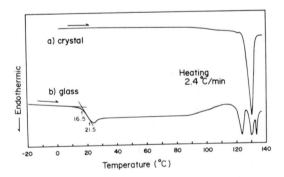

Fig. 1. Differential thermal analysis of
1,3-diphenyl-5-(p-chlorophenyl)-pyrazoline

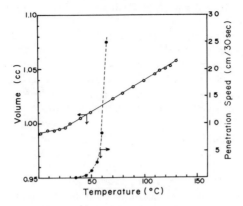

Fig.2. Dilatometric and penetrometric measurements of 1,3-diphenyl-5-(p-chlorophenyl)-pyrazoline

Scheme 1. Phase change of 1,3-diphenyl-5-(p-chlorophenyl)-pyrazoline

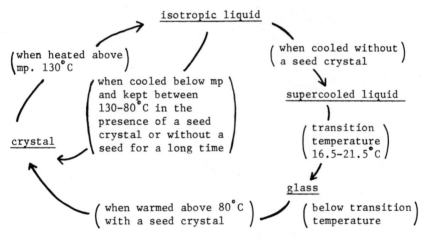

Note: The <u>crystalline state</u> is a thermodynamically stable state having a regular lattice as revealed by X-ray diffractometry. The <u>supercooled liquid state</u> is a thermodynamically equilibrium meta-stable state. The <u>glassy state</u>, which is differentiated from the supercooled liquid state, is a thermodynamically non-equilibrium state with a residual entropy, being an amorphous solid from X-ray diffractometry (4).

ZONE REFINING AND SINGLE CRYSTAL GROWTH

The zone refining of I appeared to be very difficult at first
because of its great tendency to supercool and then to change into
a stable glass as shown in the above phase change diagram. The
key point for the successful zone refining of this substance was,
as described in the EXPERIMENTAL section, to keep the cooling zone
above 80°C so as to avoid supercooling and move the zone at a very
slow rate. The single crystal growth of I was conducted in a
Bridgeman type crystal-growing furnace by means of a vessel-moving
technique. The key point for the successful growth of a large
single crystal of this substance was likewise to keep the glassy
material first at around 80°C or above for more than overnight so
as to produce seed crystals and then to melt, leaving a small
amount of the seed crystal at the bottom of the capillary. The
moving speed of the vessel, 0.2 mm/hr, was found to be a successful
operating condition under the temperature gradient of the furnace
shown in Figure 5. In this manner a single crystal of I, 1 cm in
diameter and 2 cm in length, was prepared, which had a cleavage
plane parallel to the direction of crystal growth.

The effect of the structure on the fluorescence emission
spectra of I was examined, however, no appreciable difference was
observed between in the solution, the glassy state and the single
crystal as shown in Figure 3 except the effect of the reabsorption
of the emission observed in the glassy state and especially in the
single crystal.

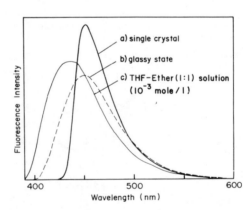

Fig. 3. Fluorescence emission spectra of
1,3-diphenyl-5-(p-chlorophenyl)-pyrazoline
measured at room temperature

EXPERIMENTAL

Materials. 1,3-Diphenyl-5-(p-chlorophenyl)-pyrazoline (I)
was prepared from p-chlorobenzaldehyde, acetophenone and phenyl-
hydrazine according to the method for the synthesis of 1,3,5-
triphenylpyrazoline described in the literature (5), and purified
by repeated recrystallizations from methanol to give pale yellow
needles, mp., 130°C.

Zone refining. In order to obtain a highly purified material
the above substance was further zone-refined with more than 50
passes of the melt-zone using a SHIMAZU Cryogenic Zone Refiner
CZ-1. The pyrazoline derivative I after repeated recrystallizations
was distilled into a pyrex glass tube of 8 mm in diameter under a
high vacuum system, and the tube was sealed off. The sealed tube
was supported in a vertical position in the zone refiner. Because
of its property to form a supercooled liquid and then a glass
very readily, the zone refining of this substance was conducted
at a very slow moving rate, 2.5 mm/hr, holding the cooling zone
above 80°C.

Dilatometry and penetrometric measurements. Dilatometry was
conducted by a conventional method. In the penetrometric
measurement a very sharp, long Japanese needle was penetrated into
the sample under the pressure of 50 g weight and its speed was
measured.

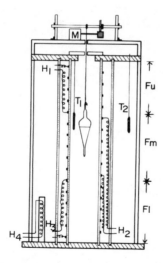

Fig.4. Bridgeman type crystal-growing furnace

  Single crystal growth. A single crystal of I was grown
from the melt in a Bridgeman type crystal-growing furnace shown
in Figure 4, which was constructed in our laboratory. A vertical furnace made of a pyrex glass tube, 2.7 cm in diameter,
had three independent temperature controle zones, being wound with
Nichrome heating coils at three separate parts (H_1, H_2 and H_3).
The appropriate temperature gradient of the furnace was obtained
by controlling the supply currents to these three heating coils.
The sample was distilled into a crystal-growing vessel of pyrex
glass under a high vacuum system (10^{-4} mmHg), and the vessel was
sealed off. The material in the vessel solidified as a supercooled
liquid, changing into a glass. The vessel was first supported at
a rather down part of the furnace and hold at around 80 C or a
little above for more than overnight to produce seed crystals,
then the vessel was hung up at the upper part of the furnace to
melt the material, only small amounts of seed crystals being left
at the bottom of the capillary, and then the vessel was moved
downward very slowly at a rate of 0.2 mm/hr under the temperature
gradient of the furnace shown in Figure 5.

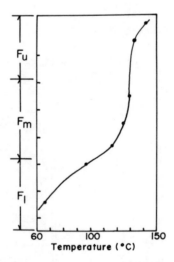

Fig. 5. Temperature gradient of the furnace
used for the single crystal growth of 1,3-
diphenyl-5-(p-chlorophenyl)-pyrazoline

REFERENCES

1. D.J.Plazek and J.H.Magill, J.Chem.Phys., $\underline{45}$, 3038 (1966).
2. S.Katsuragi, K.Okamoto, R.Matsuyama, S.Kusabayashi, and H.Mikawa, unpublished results; for pyrazoline derivatives, see Japanese Patent Application Disclosure No. 5466 of 1959, Kalle and Company, A.G.
3. For examples, G.Weiser, J.Appl.Phy., $\underline{43}$, 5028 (1972); W.D.Gill, ibid., $\underline{43}$, 5033 (1972); see also the papers by W. D. Gill and J. Mort in this volume, p. 137 and p. 127.
4. M.Sorai and S.Seki, Bull.Chem.Soc.Japan, $\underline{44}$, 2887 (1971) and references cited therein.
5. F.Bolletta and P.G.D.Marco, La ricerca scientifica, $\underline{38}$, 1062 (1968).

SEMICONDUCTIVE COORDINATION POLYMERS: DITHIOOXAMIDES COPPER COMPOUNDS

Seiichi KANDA, Asahi SUZUKI, Kuwako OHKAWA

Tokushima University, Faculty of Engineering

Tokushima, Japan

A number of coordination polymers have been synthesized in the last decade. Some of them are semiconductors[1]. One of the authors has studied the semiconductive properties of 1,6-dihydroxyphenazi-nato-Cu(II), 2,5-dihydroxy-p-benzoquinonato-Cu(II) and rubeanato-Cu(II)[2,3]. In this short note, we would like to briefly review the last one and its derivatives, to discuss on some problems in correlating the properties with the molecular structures, and to introduce the attempts to solve the problems.

In this study the substituents, R's of the N,N'-disubstituted dithiooxamides RNH-C(S)-C(S)-NHR (abbreviated as R-DTOA-H$_2$) are H-, CH$_3$-, C$_6$H$_5$CH$_2$-, C$_6$H$_{11}$-(cyclohexyl), and C$_{12}$H$_{25}$-(dodecyl). Reactions of these derivatives with copper(II) ion occur by simple mixing of the solutions of each components and the black or dark brown precipitates were obtained instantaneously. These precipitates are 1) 1:1 in molar ratio of Cu:ligand(R-DTOA-), 2) amorphous in X-ray diffraction, 3) antiferromagnetic, 4) semiconductive in temperature-resistance relation, and 5) resistors, resistivity of which slowly increases in contrast to the usual case of organic semiconductors under uniaxial pressure of 5000—15000 kg/cm^2; some of these experimental facts were reported recently[4,5,6]. The above mentioned facts, mainly 1) and 5), imply that the R-DTOA-Cu compounds have lamellar polymer structure composed from repeating unit of dimer structures of copper acetate type. A X-ray analysis[7] of dithiooxamide(rubeanic acid), NH$_2$-C(S)-C(S)-NH$_2$, and infrared spectra[8] of the derivatives suggest that these ligand molecules are stable energetically in the trans configurations. If the stability of the trans form of the ligand molecules compared with cis form is maintained even in the coordination polymers, the four tetradentate catena-μ-dithiooxamide molecules coordinate presumably to a pair of copper atoms without

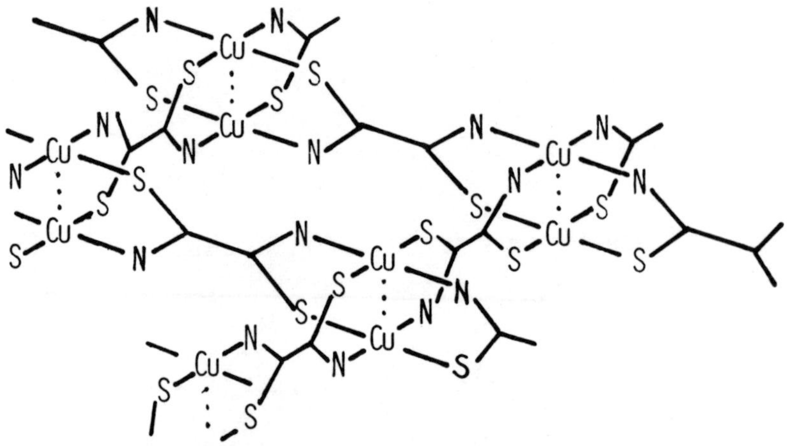

Fig.1 Non-stereospecific two-dimentional net-work model with dimer structure of copper acetate type. For simplicity, substituents on nitrogen atoms are omitted.

regularity in their relative positions. Consequently the most proba-ble structure seems to be non-stereospecific two-dimensional network on each corner of which a pair of copper ions of copper acetate type locates as shown in figure 1. The magnetic coupling parameters be-tween the paramagnetic copper ions in the two-dimentional network of the antiferromagnetic coordination polymers and the activation energies of the electronic conduction were discussed in connection with the molecular structures particularly with the substituents[6]. Because of the expected structural irregularity, however, the theo-retical interpretations are somewhat restricted. For the solution of this ambiguity stereospecific synthesis of the coordination poly-mers seems to be essential.

Surface chemistry recommends that under the condition that R is hydrophobic and the required energy for the internal rotation around the C-C axis of the central part of the R-DTOA-H_2 molecule is not extremely high, the cis form of the molecule is more stable than the trans at an air-water interface. Coordination of the cis-ligand which is spread as monolayer on water surface to copper ion supplied from the subphase (water) eliminates the above mentioned irregulari-ty for the relative positions of ligand molecules around the copper pair, and the stereospecific isomers of R-DTOA-Cu should be obtained as shown in figure 2. By this approach, the R-DTOA-H_2's with the following hydrophobic substituents, $C_6H_5CH_2-$, $C_6H_{11}-$, $C_{12}H_{25}-$,$C_{12}H_{25}$ $OCOCH_2-$, were employed as the surface active ligands in order to get the stereospecific coordination polymers. In order to check whether the expected coordination polymers are formed or not, π- A curves (surface pressure - area per molecule relation) were obtained[6].

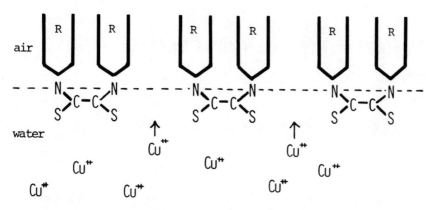

Fig.2 Supposed pattern of the reaction mechanism for stereospecific R-DTOA-Cu

The observed values of the limiting areas (area per molecule in solid state of monomolecular film) for $C_6H_5CH_2-$, $C_6H_{11}-$, and $C_{12}H_{25}-$ DTOA-H_2 and their copper compounds are much smaller than expected. That is, when the ligand molecules dissolved in the organic solvent were spread on pure water surface, surface pressure were hardly detected; when they were spread on 0.005 molar aqueous solution of cupric sulfate, the apparent limiting areas were 20, 25, and 20 $Å^2$/ molecule respectively. Since Langmuir gave the limiting area for a fatty acid with a long linear hydrocarbon chain aligned in parallel, the cross-sectional area has been estimated at about 20 $Å^2$/chain. Therefore, each molecule of $C_{12}H_{25}$-DTOA-H_2 and probably also the other two ligands and moreover these as a part of the coordination polymers should occupy more than 40 A^2 in the cis form on the water surface as far as all the added molecules exsist on the surface as monomolecular film. Likely reasons for the discrepancy might be dissolution of ligand molecules in the water and pileup of the copper compounds on the water surface.

In the case of $C_{12}H_{25}OCOCH_2$-DTOA-H_2 the observed limiting areas were 38 A^2/molecule on pure water, 51 A^2(for solid film) and >60 A^2 (for expanded film)/ molecule on 0.005 molar aqueous $CuSO_4$ solution, and these values seem to be reasonable as values in the stereospecific two-dimentional framework. If the expected polymers take the structure composed of copper acetate type dimer unit shown in figure 1, the area of a unit square shown in figure 3 is at most 60 A^2 being estimated from the bond length, and now the each square must accomodate four substituents instead of two in the case of non-stereospecific polymers as all the substituents must be on one side of the lamellar network. The resultant two-dimensional density of the substituted residues could be consistent neither with the experimentally obtained 51-60 A^2/molecule for $C_{12}H_{25}OCOCH_2$-DTOA-Cu nor the widely

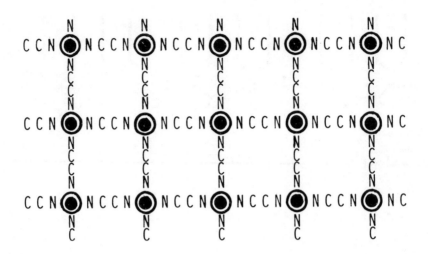

Fig.3 The top-view of the stereospecific two-dimensional net-work model with dimer structure of copper acetate type; this is not approved(see text). Double circles indicate pairs of two copper ions.

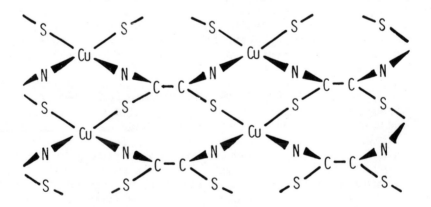

Fig. 4 The proposed molecular model of stereospecific dithiooxam-ide copper coordination polymers. Unevenness of its plane is indicated by arrowhead affixed to nitrogen atom.

approved cross-sectional area of linear hydrocarbon chain. Therefore the structure shown in figure 3 is not a satisfactory general one as the stereospecific polymers; another possible model shown in figure 4 is proposed.

These monomolecular films formed from R-DTOA-Cu polymers were laid one on top of another by the Blodgett method. So far the multi-layer films of about one hundred layers were succesfully obtained on quartz or metal plate as supporter and are being subjected to studies of semiconductive properties. The detailed results of the study will be published eleswhere.

Literature Cited

1) S.Kanda,H.A.Pohl, "Organic Semiconducting Polymers," Chapter 3, J.E.Katon, Edited, Marcel Dekker, New York, N.Y., 1968.
2) S.Kanda, S.Kawaguchi, J.Chem.Phys., 34,1070(1961).
3) S.Kanda, J.Chem.Soc. Japan, Pure Chem.Sect.,(Nippon Kagaku Zasshi), 83, 560 (1962).
4) S.Kanda, J.Chem. Soc.Japan, Ind.Chem.Sect.(Kogyo Kagaku Zasshi), 71,1426(1968).
5) H.Kobayashi, T.Haseda, E.Kanda, S.Kanda, J.Phys.Soc.Japan,18,349 (1963)
6) S.Kanda, A.Suzuki, K.Ohkawa,Ind.Eng.Chem.Prod.Res.Develop.,12,88 (1973).
7) P.J.Wheatley, J.Chem.Soc.,1965,396.
8) T.A.Scott,Jr., E.L.Wagner, J.Chem.Phys., 30,465 (1959).
B.Milligan,E.Spinner, J.M.Swan,J.Chem.Soc., 1961, 1919.

Part of this research was supported by a scientific research fund from the Ministry of Education.

GROWTH AND PURIFICATION OF PHTHALOCYANINE POLYMORPHS

P.E. Fielding

Dept. of Physical and Inorganic Chemistry,

University of New England, Armidale, N.S.W., Australia

The semiconducting properties of many organic compounds are measured with care taken to avoid spurious effects due to electroding and measuring methods, but often insufficient attention is paid to their state of chemical purity. Workers assume that extensive zone refining or sublimation procedures will yield high purity material neglecting to measure the nature, concentration and distribution of any remaining impurity. The crystal structure of the compound is, in the absence of a detailed X-ray analysis, sometimes inferred from other properties. If the organic compound can grow in different polymorphic forms under closely related growth conditions then such indirect methods can be open to misinterpretation. The aim of this discussion is to illustrate the foregoing by using as an example work carried out in our laboratory on the phthalocyanine (pc) series of compounds.

The phthalocyanines are normally prepared by methods that produce well formed acicular crystals of the monoclinic β modification. Entrainer vacuum sublimation is the best method to purify this particular form[1]. Repeated sublimations (as judged from X-ray and bulk density measurements[2]) do not improve the homogeneity of a particular batch. As well as a variation in the unit cell volume from crystal to crystal, many seemingly "perfect" crystals show a non-uniform density. Such crystals can be selected using a density gradient column[2] and when sectioned reveal the presence of minute to large, completely enclosed voids[3]. It is possible that decomposition products are trapped in these voids and one could unwittingly select such a "perfect" crystal. Even crystals of a uniform density taken from the same batch and grown in the same region of the sublimator can have a density of up to 3% less than the maximum.

We have recently published[4] the results of an investigation
carried out some years ago on the thermal stability and radiation
damage in the phthalocyanines. The results of immediate interest
concern the effect of growth conditions on the concentration of free
spins (g = 2.003) which is taken to be a measure of the impurity
content. A diamagnetic phthalocyanine such as $H_2(pc)$ with a free
spin concentration of less than 10^{15} spins/grm. was sublimed during
different experimental runs. The temperature gradient along the
sublimator tube was varied as also was the time taken to grow
crystals of a reasonable size (20 x 3 x 1 mm). The results are
shown in Table 1. In another experiment samples were sublimed at
different temperatures, the duration of each run was recorded, and
comparison samples were taken from growth regions which were at the
same temperature. See Fig. 1 for the results of typical experiments.
These series of experiments show that such purification procedures
can dramatically increase the concentration of free spins and that
relatively low free spin concentrations occur in $\beta H_2(pc)$ crystals
when the starting materials sublimed at above 500°C in nitrogen and
the crystals grown at a relatively high temperature of 450°C. We
have shown[4] that absorbed oxygen is not responsible for these
results which are thought to be due to decomposition products
trapped in the crystals. Repeated sublimations would not lead to
any increase in the purity. Varying the nature of the entrainer
gas does not appear to reduce the decomposition as Day and Price[5]
have shown that $\beta H_2(pc)$ grown in a 1:9 $N_2:H_2$ entrainer gas mixture
still contained an appreciable free radical concentration.

Barbe and Westgate[6] report trap concentrations of 10^{19}-10^{20}
trapping centers/cm^3 in $\beta H_2(pc)$, a concentration comparable to that
of the free spins reported in our work. If we accept that these
two parameters are directly related to the same impurity species
then it appears necessary to find some alternative way to prepare
single crystalline $\beta(pc)$ in which the trap concentration is much

Table 1
Sublimations performed in a nitrogen atmosphere

Subliming Temp. (°C)	Growing Temp. (°C)	Spin Concentration
600	525	1.0
	390	9.0
582	488	2.0
545	468	2.4
	360	8.7
496	420	1.6
	360	4.8
450	351	10.0

100 units equals 10^{19} spins per grm of H_2Pc

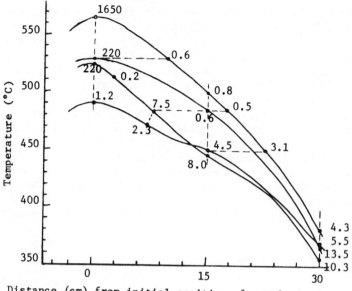

Fig. 1. Free spin line intensity from $H_2(pc)$ collected at different sublimation temperatures in a nitrogen atmosphere. Sublimation temperature 565°C, 7 g $H_2(pc)$ sublimed in 125 min; 530 and 526°C, 3 g sublimed in 150 min; 491°C, 1.3 g sublimed in 180 min. The $H_2(pc)$ showed no free spin line before sublimation. The numbers on the figure are the intensities of the free spin line per hour of sublimation per gram. A value of 100 corresponds to $(10^{19}$ spins)/g $H_2(pc)$ (i.e. 1 centre for each 100 molecules). The solid lines plot the variation of temperature with position in the furnace tube.

smaller than 10^{19}–10^{20}. Hamann[7] has reported trap concentrations for thin films of α and β $Cu(pc)$ deposited on a glass substrate as lying between 6 x 10^{14} (α) and 6 x 10^{15} (β) centers/cm^3. These polycrystalline films presumably consisted of randomly oriented crystallites similar to those reported by Lucia and Verderame[8] and differ from the epitaxial films grown by Ashida, Uyeda and Suito[9] on a mica substrate. In the latter case the crystallites grow so as to assume a definite orientation with respect to the substrate. These experimental observations indicate that oriented single crystalline films having lower trap concentrations would be more useful in electrical measurements than single crystals prepared by high temperature sublimation methods.

The literature continues to record an increasing number of polymorphic forms for the (pc) series of compounds. Recently Kirk[10] has noted in an article dealing with the crystal structure

and orientation of (pc) films that, "new structures will continue
to be found as long as new environments can be used for crystal
nucleation and growth". However, there appears to exist a great
deal of confusion as to the identity of the known polymorphs as
some authors seem to be unaware of earlier work. This concerns
particularly the relationship between forms precipitated from
sulphuric acid solutions of various concentrations and the forms
grown epitaxially on a mica substrate. The definitive work of
Honigmann and co-workers[11] has shown that two closely related
monoclinic forms of Cu(pc), αIIa and αIa are precipitated from 60%
and 70% acid solutions respectively while the triclinic αIc form
precipitates from a 96% acid solution. Ashida et al.[9] have described the preparation of oriented (pc) films grown on suitably
prepared mica substrates. Kirk[10] has questioned the assignment
ascribed to one of these films, but there seems little doubt that

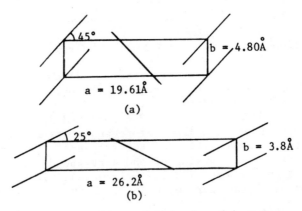

Fig. 2. ab projection of Cu(pc). (a) β form.
          (b) αIIa form.

Table 2

Honigmann's αIIa - CuPc ex 60% H_2SO_4	Ashida's b// - CuPc FILM	Honigmann's triclinic αIc - CuPc ex 96% H_2SO_4	β-CuPc
a(Å) 26.2	25.92	26.4	19.61
b(Å) 3.8	3.790	3.8	4.80
c(Å) 24.0	23.92	24.2	14.74
β(°) 93.9	90.4	90.1	121.51
γ(°) -	-	95.6	-
Z = 4	4	4	2

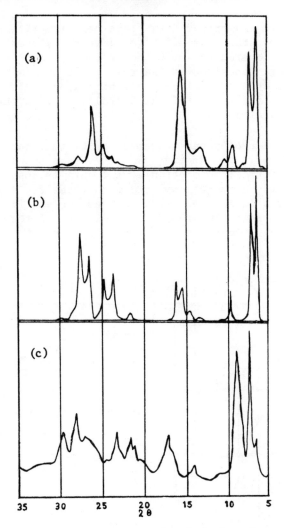

Fig. 3.   X-ray powder data for Cu(pc) polymorphs
(a)   αIIa Cu(pc).   (b)   αIa Cu(pc)
(c)   X-Cu(pc)

their b// form of Cu(pc) is identical to Honigmann's αIIa form
(see Table 2). An ab projection of the β and the αIIa forms is
shown in Fig. 2. The point to note is that the molecules in the
β form are inclined at $45°^{11}$ to the axis whereas this would be
about $25°^{11}$ in the αIIa form. It is important to note that
although there are four molecules per unit cell in the αIIa form,
they do not pack in dimeric pairs as they do in compounds such as
dianthracene and porphine.

This b$_{//}$ form was prepared by first heat treating the mica in a vacuum, (5 x 10^{-5} torr) at 300°C for 1 hour. The substrate temperature was then lowered to 150°C and Cu(pc) films of between 500Å to 900Å thick were deposited at the rate of 20-30 Å/minute. Under these conditions oriented polycrystalline regions grow in about 30 minutes such that the b axis of the Cu(pc) lies at either ±60° to the b axis of the mica.

The preparation of a new polymorph of Cu(pc), an X-form was described by Abkowitz and Sharp[12], and they gave details of X-ray diffraction and optical absorption experiments to support their claim. Their X-ray data are compared with those for Honigmann's αIIa and αIa forms in Fig. 3. These curves are sufficiently alike to suggest that perhaps the X-form is a mixture of the αIIa (or αIa) and αIc forms. In order to check this point the author

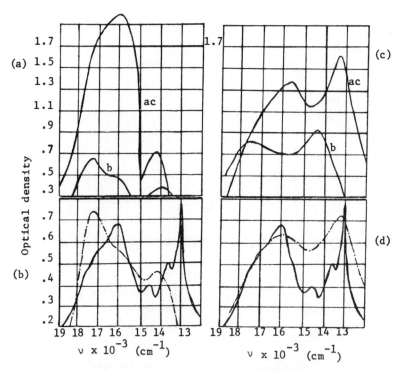

Fig. 4: Optical absorption spectra of Cu(pc) polymorphs
      (a)   ac and b polarized spectra of αIc film.
      (b)   unpolarized spectra of αIc film (–·–) and X form[12] (——).
      (c)   ac and b polarized spectra of αIIa film.
      (d)   unpolarized spectra of αIIa film (–·–) and X form[12] (——).

prepared a number of $b_{//}$ Cu(pc) films using the conditions already mentioned. It was found that, if a 500Å to 900Å thick film was deposited in 15 hours to 24 hours instead of 30 minutes a film similar in crystal structure to Honigmann's αIc form was deposited. Consequently one might expect a film to contain a mixture of these two forms, the ratio depending on the rate of deposition. The polarized and unpolarized absorption spectra of two typical films, one for the αIc and the other for the αIIa form are shown in Fig. 4. They are also compared to the unpolarized spectrum of the X-form[12]. It appears from these results that the X-form as reported by Abkowitz and Sharp consists predominately of the αIIa form with some of the αIc form present.

In conclusion I hope that I have shown that great care must be exercised in the preparation and growth of crystalline organic semi conductors in a form suitable for definitive electrical and optical measurements.

References

1. Fielding, P.E. and MacKay, A.G., Aust. J. Chem., 1964, 17, 750.
2. Fielding, P.E. and Stephenson, N.C., Aust. J. Chem., 1965, 18, 1691.
3. MacKay, A.G., Ph.D. Thesis, University of New England, 1968.
4. Boas, J.F., Fielding, P.E., and MacKay, A.G., Aust. J. Chem., in press.
5. Day, P., and Price, M.G., J. Chem. Soc.(A), 1969, 236.
6. Barbe, D.F., and Westgate, C.R., Solid State Commun., 1969, 7, 563.
7. Hamann, C., phys. stat. sol. 1968, 26, 311.
8. Lucia, A.E., and Verderame, F.D., J. Chem. Phys., 1968, 48, 2674.
9. Ashida, M., Uyeda, N., and Suito, E., Bull. chem. Soc. Japan, 1966, 39, 2616.
10. Kirk, R.S., Mol Cryst. 1968, 5, 211.
11. Honigmann, B., Lenne, H.U., and Schrödel, R., Z. Krist., 1965, 122, 185.
12. Abkowitz, M., and Sharp, J.H., J. Phys. Chem., 1973, 77, 477.

V. TRANSPORT IN AMORPHOUS STATE AND POLYMERS

MOBILITY MEASUREMENTS IN POLYMERS BY PULSED ELECTRON BEAMS

Yoshio INUISHI, Kotaku HAYASHI and Katsumi YOSHINO

Faculty of Engineering, Osaka Univ. Yamada-Kami,

Suita, Osaka, Japan

Carrier mobilities of several polymers were studied by the time of flight method using a pulsed electron beam. The obtained mobilities at the field of 1 MV/cm were 2.7 X 10^{-5} cm^2/Vsec for electron and 1.2 X 10^{-4} cm^2/Vsec for hole in PET, 1.4 X 10^{-4} cm^2/V sec for electron and 7 X 10^{-5} cm^2/Vsec for hole in PS, and 1.6 X 10^{-4} cm^2/Vsec for electron and 6.3 X 10^{-5} cm^2/Vsec for hole in PEN at room temperature respectively. Schubweg and quantum yield of the carrier were also studied. It was found that the quantum yield (number of induced carriers per impinging electron) was influenced seriously by the molecular structure.

1 INTRODUCTION

There are several methods with which the carrier mobility in dielectric materials is measured; namely photo-Hall effect[1], the time of flight method[2], surface charge decay technique[3] etc. It is not easy to obtain, however, the mobility values in polymers because of low carrier density, short lifetime and extremely low carrier mobility. Although the mobility measurements in polymers with various methods have been reported by several authors, the values of mobility are not consistent with each other[4],[5]. Recently Martin et al[6] estimated room temperature carrier mobilities in PS (polystylene) to be 1.0 X 10^{-6} cm^2/Vsec (hole), PET (polyethylene terephthalate) 1.5 X 10^{-6} cm^2/Vsec (electron) and PE (polyethylene) 4.5 X 10^{-10} cm^2/Vsec (hole) respectively using the time of flight method.

In this paper, the carrier mobilities measured by a time of flight method with electron beam, Schubweg and quantum yield in PET, PS, PEN (polyethylene naphthalate), PC (polycarbonate), PPO

(polyphenylene oxide), FEP (fluoroethylene propylene) and PTFE
(polytetrafluoroethylene) will be mentioned. The influence of
the molecular structure on quantum yield was also studied.

It turns out that our results are not in agreement with those
of Martin et al possibly due to the difference in the interpretation
of the observed transit time. It should be emphasized that to
obtain the real transit time directly from an experiment, one has
to keep "Schubweg" of the carrier to be not less than the sample
thickness[7].

2   EXPERIMENTAL METHOD

Several sorts of commercial polymers such as PET, PEN, PS, PE,
PC, PPO, FEP and PTFE were investigated. Au electrodes of 5 mm
diameter were vaccum evaporated onto both sides of each specimen.
These specimens were mounted in a vacuum cryostat inside a single
shot electron pulse beam generator. The energy and pulse width of
bombarding electron beam were $6 \sim 15$ KeV and 100 nano sec respectively. The penetration depth of bombarding electron was within $1 \sim 3 \mu$
m from the top electrode. The intensity of electron beams was
kept under about 10^6 electrons per pulse in order to minimize
the bimolecular recombination and the space charge effect. The
current produced by the induced carrier transport in the sample was
integrated with large time constant (5 msec) of the input circuit
of a preamplifier and the induced charge wave form was observed.
The repetition of the electron pulse were done after long interval
in order to minimize the space charge effect of the previous pulse.

3   RESULTS and DISCUSSION

As shown in Fig. 1, at the time t_k after the beam injection, a
clear kick was observed in the induced charge wave form in this
PET film under the application of sufficiently high bias field
($>$ 1.2 MV/cm). Furthermore, after t_k the induced charge still
increased gradually leading into the saturation. We shall call the
induced charge before t_k the "fast component" and the gradually
increasing part after t_k the "slow component". The time t_k was
proportional to the inverse of applied field E except at very high
field as shown in Fig. 2, and was considered to be the transit time
of fast carriers between electrodes. The mobility μ was estimated
from the slope of t_k vs 1/E (Fig. 2) and sample thickness d using
the relation $\mu = d/t_k E$ to be 2.7×10^{-5} cm^2/Vsec for electron and
1.2×10^{-4} cm^2/Vsec for hole respectively in PET at room temperature.

As shown in Fig. 3 electron and hole transit times t_k in PET
are also nearly proportional to the sample thickness d, and almost
independent of the beam acceleration voltage between $8 \sim 12$ kV.
Martin et al. obtained 1.5×10^{-6} cm^2/Vsec for electron mobility
by similar experiment with an assumption that only one sort of

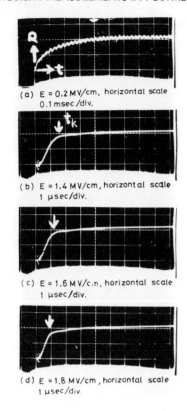

(a) E = 0.2 MV/cm, horizontal scale 0.1 msec/div.

(b) E = 1.4 MV/cm, horizontal scale 1 μsec/div.

(c) E = 1.6 MV/c.n, horizontal scale 1 μsec/div.

(d) E = 1.8 MV/cm, horizontal scale 1 μsec/div.

Fig.1 Induced charge wave form in PET by electron beam irradiation. t_k shows the carrier transit time.

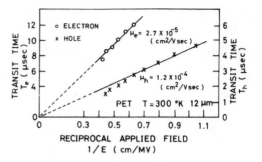

Fig.2 Dependence of the transit time on the reciprocal of the applied field.

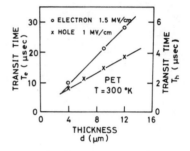

Fig. 3 Dependence of the transit time on the sample thickness.

carrier is movable in the penetration region behind to the front electrode (electron bombarding side). Because of the short penetration depth of the electron beam (a few microns), this assumption seems to give much smaller mobility than ours, although the observed transit time is similar. This assumption also can not explain the proportionality of the transit time to the thickness and weak dependence on beam acceleration voltage observed by us.

As shown in Fig.4, the collected charge obtained from the maximum value of the charge wave form vs applied field (Hecht curve)[7] saturates at the applied field above E_{th}=1.3 MV/cm for both

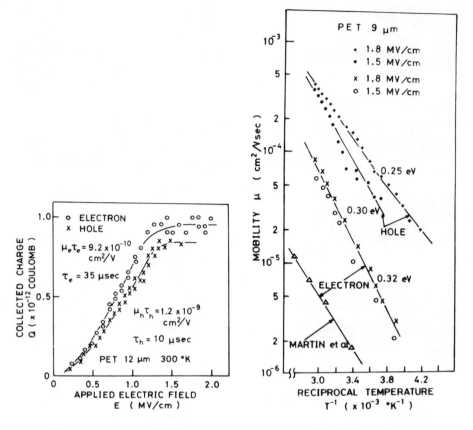

Fig. 4. Dependence of induced charge on the applied field.

Fig. 5. Dependence of the drift mobility on the inverse absolute temperature.

electron and hole at room temperature. These can be interpreted in terms of "Schubweg effect". The obtained $\mu\tau$ (mobility X lifetime) value from Hecht curves using the relation $d=\mu E_{th}\tau$ were 9.2 X 10^{-10} cm^2/V for electron and 1.2 X $10^{-9} cm^2/V$ for hole in PET at room temperature, which give the life time of $35\ \mu sec$ for electron and $10\ \mu sec$ for hole at these field respectively. The saturation electric field E_{th} decreased with decreasing sample thickness as supported from Schubweg. However, in the case of polymer, both μ and τ are supposed to be markedly field dependent, so we must be carefull on the interpretation of results. Martin et al obtained the lifetime of $0.5\,\mu$sec for electron in PET by analyzing the wave form with the assumption of the bimolecular recombination, result- ing the bimolecular recombination coefficient γ of nearly 1 X 10^{-8} cm^3/sec at room temperature. The shorter lifetime in their case may be ascribed to the larger electron beam current which enhances the bimolecular recombination. However, it should be pointed out that the above mentioned γ value is larger by several orders than the calculated γ value from Langevin equation using their mobility value[8]. When Schubweg is much smaller than the sample thickness ($\mu E\tau\ll d$), we should be careful to obtain a real transit time from a observed wave form.

In Fig. 5 semilogarithmic plots of the electron and hole mobilities in PET against the inverse absolute temperature 1/T are shown to be almost linear, indicating the activation energy of 0.32 eV for electron and 0.30 eV for hole in these temperature range at 1.5 MV/cm. The activation energy decreases slightly with increasing electric field. Much larger dependence of the activation energy on the field strength was reported by Gill[9] for PVK (poly-N- vinylcarbazole) doped with TNF (trinitrofluorenone), resembling "Poole-Frenkel" like effect.

The carrier mobilities in PS, PEN and FEP were also studied ; 1.4 X 10^{-4} cm^2/Vsec for electron and 7 X 10^{-5} cm^2/Vsec for hole in PS, 1.6 X 10^{-4} cm^2/Vsec for electron and 6.3 X 10^{-5} cm^2/Vsec for hole in PEN, and 5 X 10^{-5} cm^2/Vsec for electron and 5 X 10^{-4} cm^2/V sec for hole in FEP respectively.

The mobility in polythene obtained by Davies[3] with a surface charge decay technique are three or four orders smaller than ours. This discrepancy may be explained by taking deep trap levels into account as follows. The carrier mobility in polymer is considered to be the phonon assisted hopping mobility μ_2, which is written in the next form.

$$\mu_2 = \mu_1 \exp (- \Delta E/kT)$$

Where E is the trap depth and μ_1 is the tunneling mobility of the carrier in the localized state of the band tail. Therefore in pulse measurement at higher field, the mobilities may be determined by fast hopping electrons with relatively higher energy. However in the low field and DC measurement the mobilities may be determined by the lower energy electron near Fermi level which has much smaller hopping mobility. In fact the activation energy of the mobility obtained by Davies is several times larger than ours.

The "slow component" of charge carriers obtained by our experiment seems to be related to the fast carrier decay due to trapping into deep levels and following slow detrapping. Accordingly the application of the idea of the transit time to this component gives much smaller mobility apparently due to the inclusion of long detrapping time. For example in the case of PET the slow component apparently gives the mobility of the order of 10^{-7} cm^2/Vsec for electron or detrapping time of several hundreds μsec.

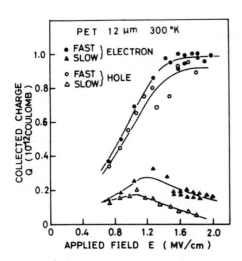

Fig.6 Field dependence of
fast and slow components of
induced charge.

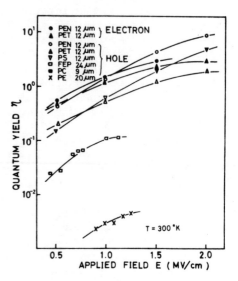

Fig.7 Field dependence of
the quantum yield of carriers
in various polymers.

    The magnitude of the slow component decreases by increasing
applied field on the contrary to the "fast component" as shown in
Fig. 6. This indicates that the slow carriers are activated to the
fast carrier by the intense electric field. It should be emphasized
again in order to obtain clear kink in charge wave form due to fast
component, we have to keep Schubweg near to sample thickness.
    In Fig. 7, the quantum yields of carriers vs applied field are
given for several polymers. The quantum yield above the saturation
field was found to be insensitive to the temperature from 250 °K
to 330 °K in PET. The quantum yield in PET, PS and PEN with aromatic nuclei was much larger as compared with that in FEP, PE, PTFE
which has not an aromatic nuclei. However, the quantum yields in
PPO and PC which have oxgen atoms just adjacent to the aromatic
nuclei were much smaller than those which has not such oxgen,
suggesting close relation between quantum yield of carriers and the
molecular structure.

REFERENCES

1)   A. I. Korr, R. A. Arndt and A. C. Damask : Phys. Rev.,
     Vol. 186 (1969) 938
2)   K. Hayashi, K. Yoshino and Y. Inuishi : Japan. J. appl.
     Phys., Vol. 12 (1973) 754 1089
3)   D. K. Davies : J. Phys. D : Appl. Phys., Vol. 5 (1972)
     162
4)   O. Dehoust : Z. angew. Phys., Vol. 27 (1969) 268
5)   R. E. McCurry and R. M. Schaffert : IBM. J., Vol. 4
     (1960) 359
6)   E. H. Martin and J. Hirsh : J. Appl. Phys., Vol. 43
7)   N. F. Mott and R. W. Gurney : "Electronic Processes in
     Ionic Crystals" Oxford Univ. Press, Oxford, England (1957)
8)   P. Langevin : Ann. Chem. Phys., Vol 28 (1903) 287 433
9)   W. D. Gill : J Appl. Phys., Vol. 43 (1972) 5033

PHOTOCONDUCTIVITY OF POLY-N-VINYLCARBAZOLE[1]

K. OKAMOTO[*], S. KUSABAYASHI[*] and H. MIKAWA[**]

* Department of Chemical Engineering, Faculty of Engineering, Yamaguchi University, Ube, Yamaguchi, 755
** Department of Applied Chemistry, Faculty of Engineering, Osaka University, Suita, Osaka 565

Many investigations have been made on the photoconductive properties of poly-N-vinylcarbazole (PVCz). However, the mechanism of
the conductivity of PVCz is still controversial. In the present
paper, some experimental results on the photoconductivities of PVCz
and the related copolymers are reported, and a possible explanation for these observations is offered.

EXPERIMENTAL

Materials.   Purified N-vinylcarbazole (VCz) was polymerized
with azobisisobutyronitril in benzene. PVCz thus obtained was purified by repeated reprecipitation. The molecular weight was determined to be 110000 by means of a membrane osmometer.

The copolymerization of VCz with some other monomers and the
purification of their copolymers were carried out in the similar
manner to those of PVCz. The content of VCz in copolymers was determined both by elementary analysis and by UV absorption measurement.

Measurements.   A thin film of polymer was formed on a nesa-
coated quartz plate by the cast method in nitrogen atmosphere.
Semitransparent main and guard electrodes were evaporated on the film
to prepare the sandwich-type cell. Several kinds of metals, such
as gold, silver, copper and aluminum, were used as the electrode
material. The cell was used to the electrical conductivity measurements without exposure to air after the evaporation of the metal
electrodes. The measurements were carried out by DC method in

vacuum (10^{-7} - 10^{-8} mmHg). A 500 W xenon lamp was used as the light source, and it was monochromatized by a grating monochromator.

RESULTS

I. Photoconductivity of Poly-N-vinylcarbazole

Figure 1 shows the voltage dependence of the photocurrent of PVCz. In applied fields lower than 7 kV/cm the photocurrent was proportional to the applied voltage. In applied fields higher than about 7 kV/cm the voltage dependence of the photocurrent, however, varied with both the polarity of the illuminated electrode and the wavelength. In the case of negative electrode illumination the photocurrent (i_{ph}^-) was proportional to the applied voltage in all the wavelength region. On the other hand, in the case of positive electrode illumination the photocurrent (i_{ph}^+) changed with the n-th power of the applied voltage (n = 1.4 - 2.0) in the UV region shorter than 360 mμ.

The photocurrent was always proportional to the light intensity (10^{12} - 10^{14} photons/cm^2 sec).

The spectral dependence of the photocurrent is shown in Fig. 2, and it varied with the polarity of the illuminated electrode, the applied voltage, and the film thickness, but it was not affected by either air or the electrode materials. In low fields no difference was observed between i_{ph}^+ and i_{ph}^- either in magnitude or in the positions of the peaks and the minima. The peaks were observed at 360, 310 and 250 mμ, and the photocurrent minima, at 330 and 290 mμ. This spectral dependence of the photocurrent corresponds inversely to the absorption spectrum except for the 250 mμ peak. This behavior is similar to that reported by Klöpffer[2]. In the applied fields higher than 35 kV/cm, a marked difference of photocurrent was observed both in magnitude and in the positions of the peaks and minima, depending on the polarity of the illuminated electrode. The spectral dependence of i_{ph}^- was similar to the case of the low applied field. However, the peaks of i_{ph}^+ were observed at 340, 295 - 300 and 240 - 250 mμ, while the photocurrent minima were observed at 315 and 285 mμ. The spectral dependence of i_{ph}^+ thus corresponded to the absorption spectrum. The magnitude of i_{ph}^+ in the UV region was 20 - 100 times larger than the corresponding value of i_{ph}^- in 35 kV/cm.

Log i_{ph} versus 1/T plots gave straight lines over the whole temperature range (20 - 160°C). In 35 kV/cm the activation energy of the photocurrent (ΔE_{ph}) was 0.07 - 0.15 eV in the visible region and 0.18 - 0.22 eV in the UV region. The latter is larger than the former by about 0.09 eV. In 150 kV/cm, ΔE_{ph} in the UV region was

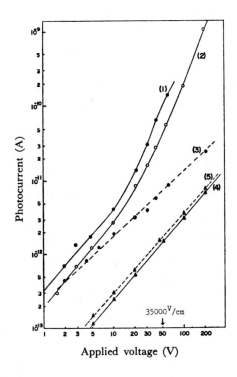

Fig. 1. Voltage dependence of photocurrent
in an Ag-PVCz-Nesa sandwich-type cell. In high
vacuum with nesa electrode illumination.
(1) peak value of $i_{ph}+$ at 330 mμ, (2) steady
state value of $i_{ph}+$ at 330 mμ, (3) steady state
value of $i_{ph}-$ at 330 mμ, (4) $i_{ph}+$ at 500 mμ,
and (5) $i_{ph}-$ at 500 mμ.

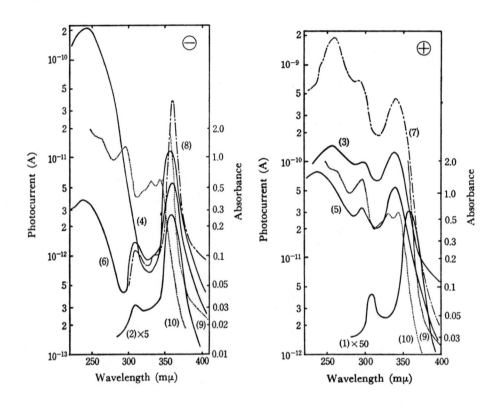

Fig. 2. Spectral dependence of photocurrent in an Au-PVCz-Nesa sandwich-type cell. Film thickness 15 μ. With positive Au electrode illumination: (1) 2 V, (3) 50 V, and (5) 50 V. With negative Au electrode illumination: (2) 2 V, (4) 50 V, and (6) 50 V. With positive nesa electrode illumination: (7) 200 V. With negative nesa electrode illumination: (8) 200 V. (1)-(4) and (7)-(8), in high vacuum. (5) and (6), in the air. (9) absorption spectrum of a PVCz film 15 μ thick. (10) absorption spectrum of a PVCz film about 1 μ thick.

0.07 - 0.10 eV, smaller by about 0.11 eV than the corresponding ΔE_{ph} value at 35 kV/cm.

II Photoconductivity of Copolymers of N-vinylcarbazole.

The comonomer used does not have any large π-electron system. The copolymer rich in VCz showed photoconductive properties similar to those observed in PVCz. As is shown in Table 1, however, the photocurrent of the copolymers was smaller in magnitude by a factor of 5 - 50 than the photocurrent of PVCz. The photocurrent decreased in all spectral region, irrespective of the polarity of the illuminated electrode. The alternate VCz fumaronitrile(FN) copolymer does not show any photoconductivity. On the other hand, the photocurrent in the PVCz film doped with polystyrene(PS) decreased by only a factor of 1.5 - 2.5 as compared with the photocurrent in an undoped PVCz film.

III Fluorescence Spectra of the Doped Films of PVCz.

In a film doped with TCNE or 1,5-diaminonaphthalene(DAN), the fluorescence of PVCz was fairly quenched and no new emission band

Table 1. Photocurrents of the copolymers of VCz

Polymer	Magnitude of the photocurrent, $\times 10^{-13}$ A/cm^2					
	with negative electrode illumination			with positive electrode illumination		
	330mμ	360mμ	400mμ	330mμ	360mμ	400mμ
PVCz	120	900	35	300	2000	80
VCz-Vinyl acetate (VAc:8 mol%)	24 (1/5)	120 (1/7.5)	7 (1/5)	—	—	—
VCz-Vinyl acetate (VAc:17 mol%)	7.7 (1/16)	58 (1/15)	3.5 (1/10)	—	—	—
VCz-Vinylpyrrolidone (VP:20 mol%)	4.4 (1/25)	42 (1/22)	0.6 (1/60)	12 (1/25)	60 (1/32)	1.6 (1/50)
VCz-Styrene (VSt:15 mol%)	17 (1/7)	130 (1/7)	2.3 (1/15)	23 (1/13)	290 (1/7)	8.0 (1/10)
PVCz doped with PS (PS:15 mol%)	56 (1/2)	540 (1/1.6)	20 (1/1.7)	160 (1/2)	800 (1/2.5)	35 (1/2.3)
Alternate copolymer of VCz and FN	0	0	0	0	0	0

Values in parentheses represents the factors of decrease in the photocurrent as compared with the photocurrent of PVCz.

was observed. In a film doped with tetramethyl-p-phenylenediamine
(TMPD), or dimethylterephthalate(DMTP), the fluorescence of PVCz
was quenched and a new broad emission band was observed at about
21000 cm^{-1}. The new broad emission band may be attributed to the
fluorescence of the exciplex.

IV Doping Effects on the Photoconductivity of PVCz.

The dopants used may be classified into three groups.

Donors (TMPD and DAN). Films doped with these donors are
characterized by the following three points; a significant decrease
in the photocurrents ($\frac{1}{100} - \frac{1}{1000}$, $i_{ph}^{+} = i_{ph}^{-}$) in all the wavelength
region, the disappearance of the superlinear dependence of i_{ph}^{+} on
the applied voltage, and a large activation energy of the photocurrent in the high-temperature range. These behaviors can be
explained in terms of the hole trapping effect of a donor. The
exciplex (PVCz^{-} --- D^{+})* does not seem to contribute to the photocurrent.

Acceptors (TCNE and DMTP). In a film doped with TCNE, a
sensitization of the photocurrent was observed in the absorption
region of PVCz as well as in the CT absorption region. Although
the doping of DMTP had no effect on the photocurrent in the visible
region, an influence was observed on the photocurrent in the absorption region of PVCz as shown in Fig. 3. The photocurrent i_{ph}^{+}
in the UV region measured under the conditions which made the space
charge effect as small as possible were fairly sensitized by doped
DMTP; there was no influence on the other properties of the photocurrent. The exciplex (PVCz^{+} --- DMTP^{-})* seems to be responsible for
the chemical sensitization of the photocurrent.

Neutral materials. Doping of Naphthalene or fluorene which
is triplet quencher in PVCz had no effect either on the photoconductivity or on the fluorescence spectrum of PVCz. In a film doped
with anthracene or perylene, the fluorescence of PVCz was quenched by
a factor of 10 to 30, while the photocurrent in the absorption
region of PVCz decreased by a factor of 2 to 7. The sensitized
photocurrent was observed in the absorption region of the dopants.

V Transfer of Electronic Excitation Energy in Polymers.

Klöpffer has reported that energy transfer in a PVCz film
occurs by exciton migration.[3] Energy transfer in copolymers was
studied by fluorescence quenching experiments using perylene and
DMTP as guest molecule. The results were tabulated in Table. 2.
As exciton migration is limited by excimer forming sites both in
PVCz and in copolymers, the number of Cz rings covered by an exciton
during the life-time are presented by $1/c_{E}$. This number is almost
the same between for PVCz and for copolymer and ranges from 400 to
900.

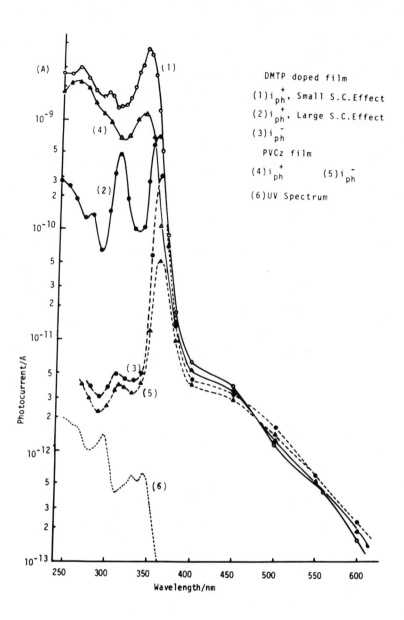

Fig. 3.   Spectral dependence of photocurrent in a sandwich-type cell of both a PVCz film and a PVCz film doped with DMTP under 35 kV/cm in a high vacuum at 20°C.

Table 2   Energy Transfer by Exciton Process in Polymer Films

No	Host	Guest	η_H	η_G	C_E x 1000	$1/C_E$
1	PVCz	Perylene	0.03	0.3	1.5	670
2	VCz-St copolymer[a]	"	0.047	0.4	1.6	630
3	VCz-VAc copolymer[b]	"	0.03	0.3	1.1	900
4	VCz-VAc copolymer[c]	"	0.035	0.35	1.9	530
5	PVCz	DMTP	0.03	0.036	2.9	350
6	VCz-St copolymer[d]	"	0.3	0.036	3.3	300

VCz mol% : (a) 79, (b) 91, (c) 83, and (d) 93.
C_E is concentration of excimer forming site, η_H and η_G are
quantum yield of host and guest fluorescence, respectively.

Discussion

I   Carrier Generation.

PVCz, a fairly strong electron donor, interacts easily even
with a very weak electron acceptor, such as DMTP, to
form an exciplex. One can expect the possibility of the existence
of impurity acting as a weak electron acceptor in a PVCz film. As
the singlet exciton in PVCz covers many carbazyl groups during its
life-time, a fairly high exciplex-forming efficiency can be expected
by the $\pi - \pi^*$ excitation. The carrier generation via the exciplex
may be considered.

$$D^* \; + \; A \longrightarrow (D^+ \text{---} A^-)^* \longrightarrow D^+ \; + \; A^- \qquad , \qquad (1)$$
$$A^- \; + \; D \; \text{or} \; D^* \longrightarrow D^- \; + \; A \qquad , \qquad (2)$$

where D is the ground state of PVCz; D^*, a singlet excited state;
A, an impurity with an electron accepting property; and A^-, an
impurity anion (immobile). D^+ and D^- are ion radicals of PVCz
(mobile charge carriers). Equation (1) shows the ionization of an
exciplex into a mobile carrier ion in a film. Equation (2) shows
the excitation of an electron donor (A^-) thermally or by a singlet
exciton. The excitations of trapped carriers and/or of acceptors by
singlet excitons may also be considered.

$$M^+ \; + \; D^* \longrightarrow M \; + \; D^+$$
$$M^- \; + \; D^* \longrightarrow M \; + \; D^- \qquad , \qquad (3)$$

where M^+ is a trapped hole or an acceptor, and M^-, a trapped electron or a donor.

It may usually be difficult for an exciplex to dissociate into
free ions in a film with a low dielectric constant. This dissociation may, however, be possible under conditions which reduce the
energy required for the separation of an ion pair,——— for example,
under a strong electric field, as has been proposed by Lyons[4].
The following two kinds of fields may be considered as strong electric fields in PVCz film: (a) the local field surrounding an ion

already produced (or a trapped carrier), and (b) the field near the surface which may act as part of a diffuse double layer.

If the processes mentioned above are supposed to be carrier generation processes in the UV region in a PVCz sandwich-type cell, the photoconductive behavior can be explained favorably as follows. In low applied fields, the exciplex will dissociate into free ions only in the local field surrounding an ion (or a trapped carrier), because the field near the surface will be small. The carriers will also be created by process (3). In a PVCz film, impurities, A, and trapped carriers (or acceptors) will exist not only at the surface layer but also in the bulk. With an increase in the absorption coefficient of a film, the illuminating light is absorbed only in the region near the surface and the favorable conditions in the homogeneous carrier generation gradually disappear. Therefore, the spectral response of the photocurrent corresponds inversely to the absorption spectrum and the photocurrent peak in the longer wavelength region shows a red shift with an increase in the film thickness, unless there is an additional condition favoring the surface generation. There is no significant difference either in magnitude or in the spectral dependence between i_{ph}^{+} and i_{ph}^{-} in low fields.

When a higher voltage is applied, the field near the surface becomes larger and may be supposed to be large enough to assist the thermal dissociation of an exciplex. With positive-electrode illumination, hole carriers generated will be swept away before recombining with the immobile A^{-} ion. A mobile electron is generated from an A^{-} ion by process (2). Thus, the thermal dissociation of an exciplex assisted by the field near the surface may act as a primary carrier generation process. Therefore, with an increase in applied voltage, significantly more carriers are generated, more favorably in the vicinity of an illuminated positive electrode rather than in the bulk. Consequently the spectral response of the photocurrent corresponds to the abosrption spectrum with positive-electrode illumination in moderate and high applied fields. During the migration in the bulk to the opposite negative electrode, the hole carriers are frequently trapped and gradually fill hole traps. As a result of the hole traps being filled, the effective mobility of a hole may become larger. With positive-electrode illumination, the number of holes reaching the negative electrode will become significantly larger with the applied voltage; i_{ph}^{+} in the UV region has, therefore, a superlinear dependence on the applied field.

On the other hand, in the case of negative-electrode illumination, most of the hole carriers generated by the same process as in positive-electrode illumination remain near the surface, without discharging at a negative electrode, and recombine with the immobile A^{-} ion. With negative-electrode illumination, the thermal dissociation of an exciplex assisted by the field near the surface seems not to act as a carrier generation process and the bulk carrier gen-

eration seems to be predominant even above a middle applied field.
The i_{ph}^{-} neither shows a superlinear dependence on the applied voltage
nor a spectral response corresponding to the abosrption spectrum,
even in high applied fields.

II Carrier Migration.

From the X-ray analysis and NMR spectra, the structure of
solid PVCz has been suggested as shown in Fig. 4. An annealed
or stretched film is crystalline and takes a pseudo-hexagonal
packing structure of a rigid rod-like molecule, with a cross sec-
tional diameter of about 12.6 Å with only a chain-to-chain order.
This may have isotactic 3/1 and syndiotactic 2/1 helix parts in a
stereoblock manner.

From the structure of solid PVCz, it seems most probable that

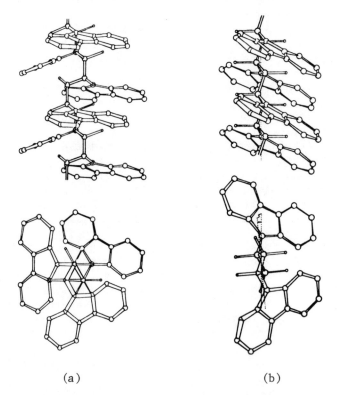

(a)   (b)

Fig. 4.   Helical structure of PVCz.
          (a) Isotactic 3/1 helix
          (b) Syndiotactic 2/1 helix

carriers can move primarily along the chain from a carbazyl group
to another adjacent carbazyl group, through the overlap of the π-
electron systems of carbazyl rings in the same polymer chain. This
carrier migration process is supported by the correlation between
the crystallinity and the photocurrent and also by the significant
decrease in the photocurrent in the copolymer of VCz with a small
amount of another monomer, which severs the overlap of the π-ele-
ctron systems of carbazyl rings in the same polymer chain, thereby
significant hindering the migration of the carriers.

REFERENCES

1) This paper is the summary of our works which have been reported
   in the following papers.
   K.Okamoto, S. Kusabayashi and H. Mikawa, Bull. Chem. Soc. Japan,
   46, 1948 (1973); ibid., 46, 1953 (1973); ibid., 46, 2324 (1973);
   ibid., 46, 2883 (1973); 46, 2613 (1973); ibid., 47, No. 1 (1974)
   in press; A. Kimura, S. Yoshimoto, Y. Akana, H. Hirata, S. Kusabayashi, H. Mikawa and N. Kasai, J. Polym. Sci. A-2, 8, 643
   (1970); S. Yoshimoto, Y. Akana, A. Kimura, H. Hirata, S. Kusabayashi and H. Mikawa, Chem. Commun. 1969, 987.
2) H. Bauser and W. Klopffer, Chem. Phys. Lett., 7, 137 (1970);
   Kolloid Z. Z. Polym, 241, 1026 (1970).
3) W. Klopffer, J. Chem. Phys., 50, 2337 (1969).
4) L. E. Lyons, "Physics and Chemistry of the Organic Solid State",
   Vol. 1, ed. by D. Fox, M. M. Labes and A. Weissberger, Inter-
   science Publishers, New York (1963) p. 392.

SENSITIZED PHOTOCURRENT IN DYE-DEPOSITED POLY-(N-VINYLCARBAZOLE)

Hiroshi KOKADO, Yutaka OKA and Eiichi INOUE

Tokyo Institute of Technology

Meguro-ku, Tokyo, JAPAN

INTRODUCTION

Photo-induced charge transfer has been studied which takes place at interface of dye and poly-(N-vinylcarbazole) (PVK). The charge transfer produces an anomalous spike in photocurrent at the early stage of illumination. Its polarity is determined solely by the type of dye employed and the magnitude is independent on the external field across the double-layer of dye and PVK. Oxygen and other electron acceptors are found to suppress the spike to appear though they assist the steady photocurrent to grow. On the basis of these findings, a mechanism of dye-sensitized photoconductivity in PVK will be discussed.

EXPERIMENTAL

On a PVK layer that was coated on Nesa-glass plates from a solution, a thin layer of dye (0.1 - 0.2 μm) was vacuum deposited. The PVK layer was about 10 μm thick. Then a 200 - 800 A silver layer was evaporated on the dye layer to serve as an electrode. A number of dyes were tried to be used but only a few were able to be actually employed due to experimental difficulties. The best results were obtained with rhodamine B, a n-type dye according to Meier's classification[1]. Hereafter description will be done mainly for this dye though pyronine G, a p-type dye will be also discussed for comparison.

Photocurrent was measured by illuminating the dye layer with monochromatic light, through the PVK layer. Unless otherwise stated, a dc voltage of 1 V was applied across the double-layer.

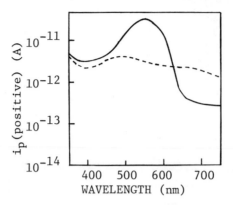

Fig.1 Spectral photocurrents in PVK layer (dashed curve)
and PVK/rhodamine B double-layer (real curve).

RESULTS AND DISCUSSION

    Dye-Sensitized Photocurrent in PVK.   A typical of spectral
photocurrents in PVK/rhodamine B double-layers is demonstrated in
Fig.1. The curve was obtained with the silver electrode positive
and the Nesa electrode negative. Here one notices clearly the
sensitization in the wavelength range of optical absorption by
rhodamine B. With the reversed polarity, however, no sensitized
photocurrent was observed. This fact suggests that the sensitized
photocurrent is carried mostly by positive holes like in single
layers of PVK. The aspect was quite the same in PVK/pyronine G
double-layers except that the sensitization was much weaker.

At longer wavelengths, a larger photocurrent is observed in the
single layer of polymer than in the dye-deposited layer. In the
single layer of polymer, there must be the photo-emission of holes
from the silver electrode to the polymer[2] that is responsible
for this effect. In other words, the dye layer between the polymer
and the electrode must retard the photo-emission.

    Photocurrent Spike.   A current spike is observed at the
beginning of photocurrent. Examples are shown in Fig.2. Here
i_p(positive) and i_p(negative) represent photocurrents observed in
specimens with the silver electrode positive and negative, respectively. Short-circuited photocurrent is represented by $i_p(0)$.
In PVK/rhodamine B double-layers, the polarity of spike is always
negative. In this case, independently upon the direction of
applied field, the current yielding the spike flows from the Nesa
to the silver electrodes. In contrast to that, deposited pyronine
G makes the spike current always flow from the silver to the Nesa
electrodes.

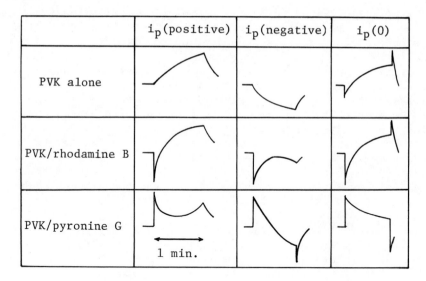

Fig.2   Illustration of photocurrent curves (not in scale)

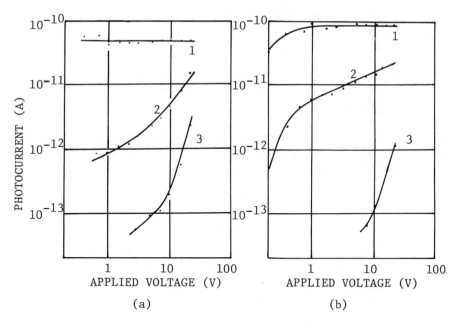

Fig.3   Initial current spike (1) and steady photocurrents (at
550 nm) in PVK/rhodamine B (2) and PVK alone (3) as function
of applied voltage.   (a) i_p(positive), (b) i_p(negative).

As is illustrated in Fig.3, the magnitude of spike is almost
unaffected by the external field across the double-layer, although
the steady photocurrent is dependent on the field.

It is apparent that there is some mechanism working which
produces a photovoltage at the interface between dye and PVK. The
most acceptable explanation is provided by assuming a charge
transfer between them. The direction of spike changes according
to the type of dye: if it is n-type, electrons are injected from
dye to PVK, and if it is p-type, vice versa. In PVK layers in
which rhodamine B is homogeneously dissolved, there is no photocurrent spike   observed, since the direction of charge transfer
is randomly oriented. A further support for the present mechanism
will be given in the next section.

Effect of Electron Acceptors.   Oxygen affects the magnitude
of spike to a great extent. The experimental results are shown in
Fig.4. Here, photocurrents were measured first in vacuum (3×10^{-4}
Torr.). During the following illumination, atmospheric air was
introduced to the measuring cell, as is indicated by arrows in

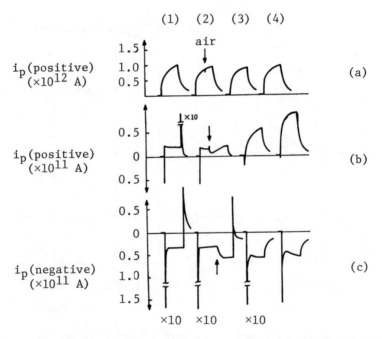

Fig.4   Influence of air on photocurrents (at 550 nm) in
PVK (a) and PVK/rhodamine B (b)(c). (1) in vacuum, (2)
exposed to air at time indicated, (3) in the air, run following (2), (4) after sufficient exposure to air

Fig.4. One should notice that photocurrents in the double-layer grew bigger by the air. In the air, the photocurrent spike became much smaller (see Curves 3) and finally disappeared (see Curves 4) except a small one in i_p(negative).

The suppression of spike by oxygen is understood as a result of trapping of photo-electrons by oxygen located at the interface. It reduces the spike by preventing the electron injection to some depth in the polymer. At the same time, the steady photocurrent increases, because the life-time of holes could be longer by less opportunity to meet electrons through their path.

When an electron acceptor is incorporated between the dye and the PVK layers, the spike is lost (anthraquinone or phthalic anhydride) or becomes very small (p-chloranil), and there is no effect of oxygen observed furthermore. On the other hand, no influence of an electron donor (diphenyl amine) is observed.

Mechanism of Dye-Sensitized Photocurrent.   A mechanism will be proposed for explaining the described photocurrent behaviors in PVK/n-type dye double-layers.

(1) $Dye + h\nu \longrightarrow Dye^*$ (excitation of dye)
(2) $Dye^* + Te \longrightarrow Dye^+ + Te^-$ (charge transfer to PVK)
(2') $Dye^* + O_2 \longrightarrow Dye^+ + O_2^-$ (charge transfer to oxygen)
(3) $Dye^+ + PVK \rightarrow Dye + PVK^+$
          $\rightarrow Dye + PVK + hole$ (hole injection)

Here, Te represents the electron trap in PVK.

The reason why Reaction (2) contributes to the current spike but Reaction (2') does not, is considered to lie in the difference of the trap distribution: Te is distributed in depth but oxygen mostly at the interface. Some potential will be needed to drive the photo-injected electrons into Te located at some distance from the dye layer, under circumstances of no external field. If there exists an internal field at the interface due to the contact barrier, it would be able to determine the direction of carrier movement, dominantly over the external field. Such an internal field has been assumed by Salaneck in the case of contact of PVK to CdS and CdSe[3]. The spike current is supposed to compensate the internal potential and as the result facilitate Reaction (3) to occur.

REFERENCES

1. H. Meier, Spectral Sensitization (Focal Press, 1969), p.166.
2. A. I. Lakatos and J. Mort, Phys. Rev. Lett. 21 1444 (1968).
3. W. R. Salaneck, Appl. Phys. Lett. 22 11 (1973).

CHARGE INJECTION AND TRANSPORT IN DISORDERED ORGANIC SOLIDS

J. Mort

Xerox Webster Research Center

Webster, New York   14580

I.   INTRODUCTION

In the development of photoreceptors for use in commercial imaging systems, such as electrophotography, an important practical requirement is the ability to fabricate large area, defect-free films capable of maintaining their mechanical integrity under a wide range of operational conditions. Coupled with this are the stringent specifications regarding the photoelectronic properties of the material if it is to prove acceptable for incorporation into a particular device. A very successful class of materials which fulfill many of these requirements are the inorganic amorphous chalcogenides. In the realm of organic solids the analogous class of materials are polymeric solids since large area polymeric films can be readily made from solution. Consequently there is currently great interest in the investigation of the photoelectronic properties of polymers and the generic state of which they are a particularly important class, namely the amorphous organic solid state.

With this background, the purpose of this paper is to discuss experimental studies of two polymers, poly(N-vinyl carbazole), PVK, and the charge transfer complex between poly(N-vinyl carbazole) and trinitrofluorenone, PVK-TNF. More recently the studies have been extended to vacuum evaporated trinitrofluorenone, TNF, films and the most significant results of these measurements will be summarized.

II. EXPERIMENTAL DETAILS

The experimental techniques involve the injection of carriers photogenerated in thin sensitizing layers of amorphous selenium which are in contiguous contact with much thicker transport layers of the material under study. The details of the photoinjection and subsequent motion of charge across the transport layers are determined by either the xerographic discharge technique or the time of flight method. Both the techniques and details of sample preparation have been described in detail elsewhere [1,2] and only the general principles will be outlined here. The polymer films ranging in thickness from 10μm to 20μm are coated onto conductive substrates from polymer solutions. After heat treatment to remove residual solvents a much thinner layer, ∿1μm of amorphous selenium, is vacuum evaporated onto the polymer film which is held at 55°C during the selenium evaporation. This completes the sample preparation steps for the xerographic discharge studies since the corona charging can be conveniently accomplished on the free selenium surface. For the time of flight investigations a further evaporation of a semitransparent gold electrode is required to permit the application of a constant voltage source.

The time of flight method permits a direct measurement of carrier drift velocities by measuring the time a thin sheet of charge photogenerated in the selenium takes to drift across the total structure. In the xerographic discharge technique the rate of discharge of the corona-produced surface potential as a function of time (and therefore voltage) is monitored by means of a capacitively coupled probe. By comparison with theoretical predictions [3,4] it is possible to deduce information such as injection efficiencies and, under appropriate experimental conditions, carrier drift velocities.

III. EXPERIMENTAL RESULTS

(a) Poly(N-vinyl carbazole)

Figure 1. indicates a typical current pulse obtained in time of flight studies resulting from the drift of photoinjected holes across a PVK film.[1] Several unusual features are apparent in this current pulse. In the first place, despite the fact that a very narrow sheet of charge was injected into the PVK sample the carriers exhibit a wide range of arrival times at the back face of the sample. The arrival times are so dispersive that it is difficult to define a unique transit time although operationally a time t_T is defined by the shoulder in the current pulse.

This roughly defines the transit time of the fastest carriers. A second unusual feature is that the inverse of this time and hence the carrier velocity is a superlinear function of field. The ques-

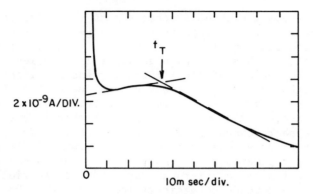

Fig. 1.   Transit pulse of holes photoinjected from Se into PVK.
          After Mort [1].

tion naturally arises as to whether one can meaningfully define a
carrier mobility under such conditions.   Remembering that mobility
is defined as the mean velocity of all carriers per unit field, one
is faced with only two possibilities.   The first is that a distribution of mobilities exists for carriers in PVK, or second, that the
current pulse displays directly the velocity distribution of the
carriers and so no mean velocity exists, on this time scale, for the
carriers.   This latter statement is equivalent to stating that the
charge transport occurs under completely non-equilibrium conditions.
The possibility that a distribution of mobilities exists in PVK
requires that many different and discrete transport paths exist
within the polymer.   Although remotely possible, the observation of
similar transport phenomena in a range of different materials such
as PVK:TNF,[5], TNF,[6] and As_2Se_3[7] make this an unlikely interpretation.   For a more detailed discussion of these points and a
theoretical model, which may account for these features, based on
the stochastic nature of transport in disordered solids, the reader
is directed to the recent work of Scher.[8]   In the past it has been
customary to operationally define mobilities for holes in PVK by use
of the fastest transit time.[1,9]   This approach led to mobilities
$\sim 10^{-6}$ cm^2/Vsec at fields of 10^5V/cm and room temperature.   In view
of the recent theoretical advances by Scher,[8,10]   it may be more
exact to characterize carrier transport in solids such as PVK in
terms of the extraordinarily low carrier velocities observed in
comparison with carrier velocities attained at the same fields in
crystalline solids.

Figure 2 shows a plot of discharge rate versus field E where
E = V(t)/L obtained from xerographic discharge studies of Se:PVK
structures for a range of light intensities. The important feature
of these results is that at very high light intensities the discharge
rate becomes independent of light intensity and the light-intensity-
independent discharge rate is a very strong function of E. This has
previously been interpreted in terms of achieving a space-charge-
limited condition; [1,2,3] i.e. an amount of charge is injected in
PVK (in a time short compared with the transit time of the leading
edge of the injected sheet) which equals the initial surface charge.
This space-charge-limited curve thus allows a determination of the
velocity of the fastest carriers at any value of the field. The
very strong field dependence again reflects the fact that the
carrier velocities are a superlinear function of field. Analysis
of the injection current levels achieved are interpreted as
meaning that any energy barrier existing between the transport
states for holes in amorphous selenium and PVK is not a limiting

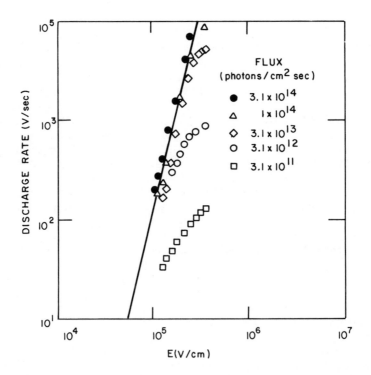

Fig. 2. V̇ versus E = V(t)/L for a Se:PVK structure at five different light intensities. After Mort et al [2].

factor in determining the discharge rate. Recent studies by
Nielsen [11] confirm that the energy barrier is ∿0.3eV and that,
in the absence of interfacial trapping, this would not be a limitation on the time scale of transit times through PVK.

(b) Poly(N-vinyl carbazole) - Trinitrofluorenone

This weak charge-transfer complex has been the subject of
considerable recent study. Schaffert has described the electro-
photographic properties of this organic photoreceptor material [12]
and Gill has recently reported [5] (see also page 124 of this
volume) extensive time of flight studies. As in the case of PVK
the studies to be described here deal with the injection of photogenerated carriers from amorphous Se into films of PVK:TNF. In
particular the injection studies have been carried out on PVK:TNF
films of variable TNF loading. The details of these experiments,
have been described elsewhere [13]. Unlike the case of PVK, both
electron and hole injection has been studied by use of either
positive or negative corona charge in the xerographic discharge
technique. As described in the paper by Mort and Emerald, [13]
a very large asymmetry in discharge rate was observed at low TNF
loadings for the two corona charge polarities; the discharge rate
for negative corona charging, corresponding to electron injection
into and transport across the PVK:TNF, was orders of magnitude
smaller than for positive corona charge. As the TNF loading was
increased the asymmetry decreased such that at about 0.2 molar
ratio the discharge rates were approximately equal. As a working
hypothesis it was assumed that the transport states for both
electrons and holes in amorphous Se were essentially aligned with
the corresponding states in PVK:TNF. Consequently the asymmetric
discharge rates were assumed not to reflect different injection
efficiencies but simply a manifestation of widely different electron and hole velocities at the same field, resulting in different
space-charge-limited xerographic discharge rates. In this way operational carrier mobilities for electrons and holes at a fixed field
were determined from the space-charge-limited xerographic discharge
curves as a function of TNF loading. Figure 3 shows a plot of
these results. In Figure 3 the mobilities determined by Gill [5]
by direct time of flight studies are also plotted. Both results
show the strong increase of   electron mobility and decrease of
hole mobility with increased TNF loading. The agreement between
the two independent techniques was taken by Mort and Emerald [13]
as evidence that the original hypothesis concerning the alignment
of transport states in amorphous Se and PVK:TNF was correct. Since
the energy gap between transport states for electrons and holes in
amorphous Se is known to be ∿2.1eV,[14] this allows the conclusion
to be drawn that the transport states in PVK:TNF are separated by
∿ <2.1eV. This is an important advance in our understanding of

the nature of the charge-transfer complex process. It suggests
that the electron transport in the complex occurs through the
essentially unperturbed lowest empty molecular orbital of TNF and
the hole transport occurs through the essentially unperturbed
highest occupied molecular orbital of PVK. A corollary of the
model is that the highest occupied molecular orbital of TNF should
lie significantly (∿1-2eV) below that of PVK.

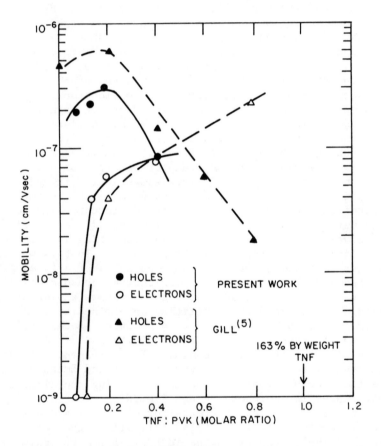

Fig. 3.   Electron and hole mobilities in PVK:TNF as a function of
          TNF loading. After Mort and Emerald [13]. The triangles
          represent the data of Gill [5].

(c) Trinitrofluorenone

   The observation of very efficient injection from amorphous Se
into films of PVK:TNF together with the implication that electron
transport in the complex occurs through the essentially unperturbed
lowest empty molecular orbital of TNF leads to the expectation
that efficient electron injection should be observed from amorphous
Se into amorphous films of TNF itself.   Recent experiments [15]
show that this expectation is realized.   Amorphous films of TNF can
be prepared by vacuum evaporation.   Although initially amorphous,
the films rather rapidly crystallize in a few hours into the poly
crystalline state.   In order to study the films in the amorphous
state the amorphous TNF films ∿10μm thick were evaporated onto
previously evaporated thin sensitizing layers of amorphous Se
∿1μm thick.   In this way the structures were studied using the

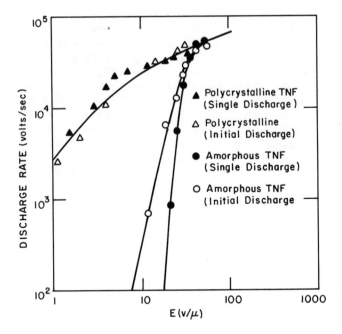

Figure 4.   V̇ versus E = V(t)/L for Se:TNF structures.   The circles
and triangles correspond to amorphous and polycrystalline TNF respectively.   After Emerald and Mort [15].

xerographic discharge technique within a few minutes of completion of the TNF evaporation. Figure 4 shows the results of the xerographic discharge rate measurements on a structure of Se:TNF where the TNF was amorphous and measurements on the same sample after the TNF had converted to the polycrystalline phase. For both samples the xerographic discharge was measured by analysis of a single discharge curve and by the initial discharge rate method. As discussed elsewhere [2] the observation of lower discharge rates at low fields determined from a single discharge curve is direct evidence for the occurrence of deep trapping due to a range limitation. Focusing attention on the initial discharge rate data it is seen that the discharge rates obtained for the polycrystalline TNF exhibit a much weaker field dependence than obtained when the TNF is amorphous. As discussed by Emerald and Mort [15] the discharge rates for amorphous TNF are interpreted as arising from the achievement of a space-charge-limited discharge at the very high flux used, $\sim3 \times 10^{14}$ photons/cm^2 sec. As in the studies on PVK and PVK:TNF, it is possible to determine the highest carrier velocities from this curve and the steep field dependence again indicates that the carrier velocity is a superlinear function of field. In terms of an operationally defined mobility this gives a value of 10^{-5} cm^2/Vsec at 10^5 V/cm. The discharge rate observed for polycrystalline TNF has the characteristic field dependence of the xerographic discharge rate observed for amorphous Se. This implies that the limiting step in the discharge of the Se:polycrystalline TNF is the rate of photogeneration in the amorphous Se and the velocity or mobility of electrons in polycrystalline TNF, unlike the case of amorphous TNF, does not limit the discharge rate. It is therefore concluded that a very large increase in carrier velocities (for a given field) occurs when the TNF film crystallizes. In order to account for the observed change in discharge rates this increase in carrier velocities amounts to several orders of magnitude. It can be concluded therefore that even in a molecular solid the disorder of the amorphous state has a drastic effect on the transport of charge.

ACKNOWLEDGEMENTS

The author would like to acknowledge the collaboration of R. L. Emerald in many of the experiments described in this paper, particularly those involving the studies of PVK:TNF and TNF films.

REFERENCES

1. J. Mort, Phys. Rev. B5, 3329 (1972).
2. J. Mort, I. Chen, R. L. Emerald and J. H. Sharp, J. Appl. Phys. 43, 2285 (1972).
3. I. Chen and J. Mort, J. Appl. Phys. 43, 1164 (1972).

4.   I. Chen, R. L. Emerald and J. Mort, J. Appl. Phys. $\underline{44}$, 3490
     (1973).
5.   W. D. Gill, J. Appl. Phys. $\underline{43}$, 5033 (1972).
6.   W. D. Gill, Proc. V[th] Int. Conf. Amorphous and Liquid Semicond.
     Garmisch-Partenkirchen 1973 (to be published).
7.   M. E. Scharfe, Phys. Rev. $\underline{B2}$, 5025 (1970).
8.   H. Scher, Proc. V[th] Int. Conf. Amorphous and Liquid Semicond.,
     Garmisch-Partenkirchen 1973 (to be published).
9.   P. J. Regensburger, Photochem. and Photobiol. $\underline{8}$, 429 (1968).
10. H. Scher, Phys. Rev. (to be published).
11. P. Nielsen, Photo.Sci. and Eng. (to be published).
12. R. M. Schaffert, IBM J. Res. Develop. $\underline{15}$, 75 (1971).
13. J. Mort and R. L. Emerald, J. Appl. Phys. (to be published).
14. J. Mort and A. I. Lakatos, J. Non-Crys. Solids $\underline{4}$, 117 (1970).
15. R. L. Emerald and J. Mort, J. Appl. Phys. (to be published).

CHARGE TRANSPORT IN TNF:PVK AND TNF:POLYESTER FILMS AND IN LIQUID, AMORPHOUS AND CRYSTALLINE TNF

W. D. Gill

IBM Research Laboratory

San Jose, California 95193

INTRODUCTION

Charge transport of excess photoinjected carriers has been studied in a number of materials containing 2,4,7 trinitro-9-fluorenone (TNF). Initially measurements were made on charge transfer complexes of TNF with poly-N-vinylcarbazole (PVK). The effect on carrier mobilities of systematic variations of material composition was used to investigate the details of the microscopic transport mechanisms.[1] Similar studies have now been made on the somewhat simpler system in which TNF is dispersed in a polyester matrix. This system has the advantage that no charge transfer complexing occurs and also much higher TNF concentration samples can be prepared. The studies of these polymer systems showed that electron transport is due to intermolecular hopping through TNF states. We have recently extended this study of electron transport in TNF to examine the effects of structure on mobility in the pure TNF system. Drift mobility measurements have been made on liquid, amorphous and crystalline states of TNF. This is the first organic material in which mobility has been measured in all three states.

TNF:PVK AND TNF:POLYESTER FILMS

The TNF:PVK system, ranging in composition from pure PVK to 1:1 molar ratios with respect to the monomer units, has been most extensively studied. Optical absorption and electroabsorption,[2] carrier generation,[3] transport,[1] and dielectric properties[4] of these films have been investigated. Both hole and electron drift mobilities measured by time-of-flight techniques were observed for all compositions of the mixed TNF:PVK materials. With pure PVK and pure TNF

only hole mobility and electron mobility respectively could be determined. The magnitudes of the drift mobilities were extremely small
and were strongly electric-field and temperature dependent. Both
electron and hole mobilities showed the same field dependence and
both had zero-field activation energies near 0.7 eV for all film
compositions. The hole mobility was greatest in pure PVK and decreased rapidly with increasing TNF. For electrons no transport could be
detected in pure PVK films but as TNF was added electron mobility
was measured and increased sharply with TNF concentration.

The low mobilities and their strong concentration and temperature dependences suggest that transport of both carriers are thermally assisted intermolecular hopping processes. The composition
dependence of hole mobility is in good agreement with a model in
which hopping between uncomplexed PVK monomer units is considered.
If this uncomplexed PVK concentration is expressed in terms of an
average intermolecular separation R, an exponential dependence of
μ/R^2 versus R is observed which can be attributed to changes in the
overlap of the localized hole states. From this exponential dependence we obtain a value for the hole localization parameter α equal
to 0.91 X 10^8 cm^{-1}.

Since the densities of both the uncomplexed TNF and the total
TNF increase rapidly with increasing TNF concentration, the TNF:PVK
data alone was not sufficient to distinguish which TNF density was
active in electron transport. This question was resolved by making
drift mobility measurements over a range of concentrations of TNF
dispersed in a non-complexing polyester matrix. The electron mobilities were found to have the same magnitudes and field dependence as
was measured for comparable total TNF concentrations in the TNF:PVK
system. This result established that in TNF:PVK all the TNF molecules whether complexed or uncomplexed are active in the hopping
process. At low TNF concentrations an exponential dependence of
μ/R^2 versus R, where R is the average separation of TNF molecules,
is evidence for intermolecular hopping transport of localized electrons. The localization parameter for electrons is $\alpha = 0.56$ X 10^8
cm^{-1}.

At the higher TNF concentrations, i.e. where TNF is the major
constituent, the electron mobility increases more rapidly than
exponentially with decreasing intermolecular distance. In quenched
films of pure TNF the mobility activation energy has decreased to
0.63 eV from about 0.7 eV in the low TNF concentration material.
However this decreased activation energy does not account for the
entire increase in mobility. Since deviation from exponential
behavior becomes significant at intermolecular separations approaching the maximum dimensions of the TNF molecule it is very likely
that more complex intermolecular interactions play a role at the
higher concentrations.

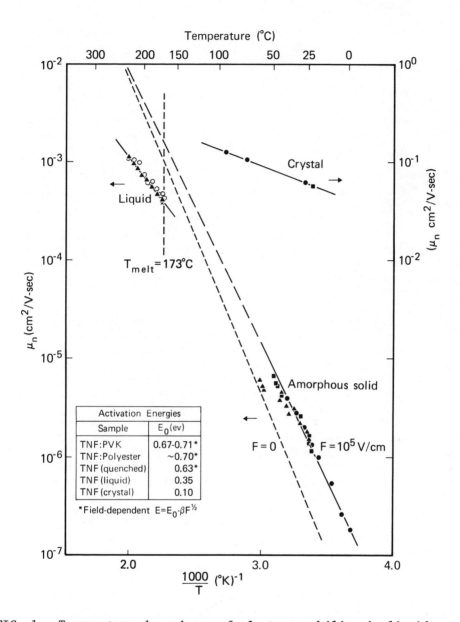

FIG. 1. Temperature dependence of electron mobility in liquid, amorphous solid and crystalline forms of pure TNF. The dashed line is the zero-field mobility for the amorphous solid. The insert shows the mobility activation energies for the pure TNF and for TNF in PVK and in polyester.

LIQUID, AMORPHOUS AND CRYSTALLINE TNF

Electron transport studies in TNF have been extended to examine the effects of structure through mobility measurements on amorphous, liquid and single crystal samples.[5] The results are shown in Fig. 1. In amorphous TNF formed by quenching from the melt, the electron mobility was field dependent with very similar characteristics to those observed in the polymer films described above. In liquid TNF well-defined transients corresponding to electrons were observed. The mobilities were field independent and the temperature dependence indicated an activation energy of 0.35 eV. In crystalline TNF electron mobility was measured in thin single crystal platelets. The mobility was field independent with a magnitude of 0.06 cm^2/V-sec at 22° C. The temperature dependence indicates an activation energy of about 0.10 eV.

For crystalline TNF, the magnitude of the mobility is similar to that observed in many other molecular crystals. The activated behavior and fairly low mobility suggest hopping, possibly by small polarons, as the transport mechanism. However a trap-controlled process cannot be excluded.

In the liquid and amorphous material the experimental evidence is very strong for transport by hopping of highly localized electrons. The small magnitudes of the mobilities, the concentration dependence in the TNF:polymer films and the reproducibility for many samples support this model. Experiments comparing the mobilities in the liquid and amorphous state for pure TNF and TNF:polyester have established that the carriers observed in the liquids are the same as those observed in the solid films being electrons and not ions. The main difference in the transport behavior of the liquids and of the amorphous solid materials is the disappearance in the liquid state of the field dependent part of the activation energy. Continuous measurement of the mobility in going from the amorphous solid to the liquid has not been possible due to sample crystallization above 40° C. However if such data could be obtained a smooth transition would be expected since only a decrease in viscosity should occur.

SUMMARY

Drift mobilities have been investigated in TNF:PVK and TNF:polyester films over a wide composition range. The mobilities all have small magnitudes and the electric field and temperature dependences show all compositions to have large field dependent activation energies. Concentration dependence has shown that transport of both holes and electrons are by intermolecular hopping processes of highly localized carriers. Hole transport is associated with carbazole states while electron transport is associated with TNF states.

Comparison of mobility data for TNF:PVK and TNF:polyester has established that electron transport is unaffected by charge transfer complexing of TNF with carbazole.

Mobilities have also been studied for pure TNF in the liquid, amorphous solid and crystalline states. In single crystals the magnitude of the mobility is fairly typical of values observed in other molecular crystals being 0.06 cm^2/V-sec at 22° C. The transport mechanism is by hopping of small polarons or possibly a trap-controlled drift process. In the disordered states of TNF the mobility is several orders of magnitude smaller and is strongly activated. The mobility in the amorphous solid is field dependent however the field dependent part of the activation energy disappears in going to the liquid state. Transport in the disordered TNF is by intermolecular hopping of highly localized electrons.

REFERENCES

1. W. D. Gill, J. Appl. Phys. 43, 5033 (1972).
2. G. Weiser, J. Appl. Phys. 43, 5028 (1972); Phys. Stat. Sol. (a) 18, 347 (1973).
3. K. K. Kanazawa, Bull. Am. Phys. Soc. 18, 451 (1973).
4. B. H. Schechtman, Bull. Am. Phys. Soc. 18, 451 (1973).
5. W. D. Gill, to be published in the Proceedings of the 5th Int. Conf. on Amorphous and Liquid Semiconductors, Garmisch-Partenkirchen, September 1973.

VI. SUPERCONDUCTIVITY

THE PROBLEM OF SUPERCONDUCTIVITY IN ORGANIC AND ORGANO-METALLIC COMPOUNDS

W. A. Little

Stanford University

Stanford, California

I would like to give a brief introduction to the subject of superconductivity in preparation for the panel discussion. I hope that in this way the two segments at this conference can be brought together - those who have been working principally with organic semiconductors, and those who are working in superconductivity.

Superconductivity is a rather dramatic phenomenon. This can be seen from measurements of the electrical resistance of a super-conducting metal as a function of temperature in Fig. 1. As one cools the metal down, the resistance drops smoothly, begins to flatten out and then suddenly drops precipitously to zero. At temperatures below this critical temperature, T_c, the resistance is found to be essentially zero.

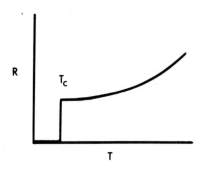

Fig. 1

The phenomenon is not rare. Twenty six elements are known to become superconducting, ten more do so under high pressure, and over two thousand superconducting alloys are known. It is thus very much more common than ferromagnetism.

In Fig. 2 we show a histogram of the distribution of the number of known alloys as a function of transition temperature. The highest known transition temperature for any alloy is only 21K and hence sophisticated refrigeration techniques are needed to use or to observe superconductivity.

For economic and scientific reasons much work has been done to attempt to raise the transition temperature of materials. The progress which has been made to date is illustrated in Fig. 3.

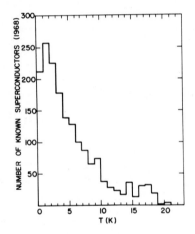

Fig. 2

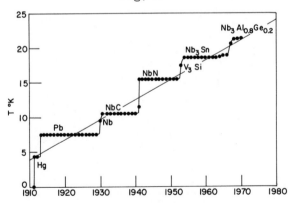

Fig. 3

This suggests that it is unlikely that one will find an alloy with
a transition temperature substantially above 21K, at least in the
next decade. One may well ask, "How high a T_c can one obtain?"
To answer this we need to know something about the theory of superconductivity.

Our present understanding of superconductivity is based on the
theory of Bardeen, Cooper and Schrieffer (BCS) put forward in 1957.[1]
In this they postulated that the electrons in the metal are not free
as in normal metal but rather are bound to one another in pairs;
and secondly, that these pairs form a Bose-Einstein-like condensate
such that all the pairs are in the same quantum state. The attractive interaction which is needed to form the pairs arises from the
interaction with the lattice. This is illustrated in Fig. 4. As
an electron moves through the lattice it interacts with the positive
ions of the lattice. The ions move towards the instantaneous
position of the electron and produces a deformation in the lattice.
In this deformed part of the lattice there is thus a slight excess
of positive charge and a second electron is attracted to it.

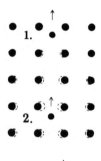

Fig. 4

This mechanical model singles out just one pair of electrons.
However, because the other pairs are in the same state they can
only interfere constructively with the first pair and thus the
effective binding of the pairs is amplified. At very low temperatures essentially all the electrons are bound as pairs, but, as one
raises the temperature some pairs break up and now interfere
destructively with the others. At higher temperatures this becomes
catastrophic and above a critical temperature T_c no pairs remain and
one has the "normal" state of the metal.

The detailed microscopic theory shows that the transition
temperature is given by the expression,

$$kT_c \approx \hbar\omega \exp\left\{\frac{-1}{N(0)(V_{ee} - V_c)}\right\} \qquad (1)$$

where ω is the Debye frequency of the lattice, V_{ee} is the lattice or phonon induced electron-electron interaction and V_c is the Coulomb repulsion.

The form of this expression suggests that one might be able to obtain a high T_c if one could make the lattice more polarizable, i.e. if the lattice restoring force was made weaker. In this case the electrons could deform the lattice more strongly and thus pro- duce a greater excess charge which would lead to greater binding. This is partially true as can be seen in Fig. 5 where we show the variation of T_c with the inverse of the force constant, k, of the lattice.

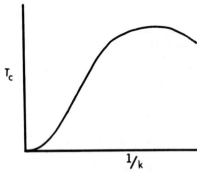

Fig. 5

However, one sees that T_c does not rise indefinitely, but rather flattens out, reaches a maximum and then decreases. The reason for this is that as one softens the lattice, the Debye frequency ω in the pre-exponential factor of (1) also becomes smaller and eventually dominates the expression. A second factor arises through the electron-phonon interaction. As one makes the intrinsic lattice force constant weaker the electron-phonon inter- action itself begins to contribute to the net force constant. This results in a renormalization of the phonon frequencies to lower values and eventually to a lattice instability. These factors place a natural limit on the transition temperature one can obtain using the phonon mechanism.

To avoid these limitations we have proposed[2] using instead of the phonon interaction resulting from the lattice polarization an interaction resulting from an electronic polarization. By so doing the Debye frequency in (1) is replaced by an electronic excitation energy and provided one can get a sufficiently strong V_{ee} one should obtain substantially higher transition temperatures. This new mechanism has been called the excitonic mechanism of superconductivity.

We proposed originally that such a system might be built up from a conjugated organic polymer as the conductive "spine," with polar-izable dye-like molecules attached to it. This was a somewhat impractical example but in recent years a number of promising structures have been found which appear capable of satisfying the theoretical criteria.

One possibility is the class of compounds studied by Krogmann[3] in which a chain-like structure is formed between platinum atoms surrounded by a square planar structure of organic ligands. The ligands can be replaced with highly polarizable moieties. Another is the class of TCNQ charge transfer compounds which form linear stacked arrays.[4] These two types are illustrated in Fig. 6.

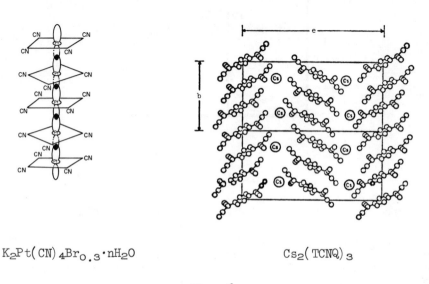

$$K_2Pt(CN)_4Br_{0.3}\cdot nH_2O \qquad\qquad Cs_2(TCNQ)_3$$

Fig. 6

In all these materials the conductive spine is essentially one-dimensional, and this introduces some special problems. As indicated earlier the pairs must condense into the same state, however, the nature of this state can vary. This can be described by an equation for the Free energy, $F(\psi)$ obtained by Ginzburg and Landau:

$$F(\psi) = a\int|\psi|^2 d\tau + \frac{b}{2}\int|\psi|^4 d\tau + c\int\psi(\nabla - \frac{eA}{c})^2\ \psi d\tau$$

where $|\psi|^2$ represents the density of the condensed pairs, and a, b, and c are constants. The probability of finding a particular form for ψ is then given by $\rho|\psi| = \exp -F(\psi)/kT$, and the system fluctuates

among these different configurations. In a one dimensional system
the constraints on ψ are much weaker than in three dimensions, with
the result that fluctuation effects are much bigger. One finds
then that the transition instead of being sharp as it is in three
dimensional systems, is broadened as shown in Fig. 7. One finds
that the resistance begins to drop at temperatures well above the
mean field, T_c but only goes to zero at a temperature well below
this T_c.

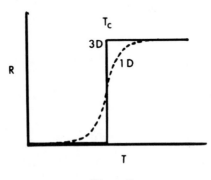

Fig. 7

A second difficulty arises because of an instability of the
lattice - the, so-called, Peierls' instability. If one has a uniform chain of atoms with one electron per atom then the resulting
conduction band will be half filled. One can see that such a system
must be unstable at T = 0 K, for if it should distort so as to
double the size of the unit cell, this would put a gap in the band
at the Fermi surface. The filled states below the band edge would
thus be lowered in energy while the empty states above the gap
would be raised in energy. A net gain in energy would thus result
leading to a spontaneous distortion. This is equivalent to a Jahn-
Teller distortion in the crystal as a whole. At higher temperatures
the distortion need not occur though, for as one raises the temperature some of the states above the gap become occupied and some of
the states below become emptied so the energy gained by distorting
the lattice is reduced. Above a critical "Peierls'" temperature,
T_p the undistorted lattice is favored.

Just as the thermal occupation of the levels above the gap
tend to suppress the Peierls' transition so electron-electron interactions also suppress the Peierls' transition because these interactions mix in states below the gap with those above the gap and

the energy gain from the distortion of the lattice is consequently reduced.[2] As a result the Peierls' transition and the superconducting transition temperatures must be calculated in a self-consistent manner in the one dimensional system.[6]

Thirdly, just as fluctuations depress the transition to a fully superconducting state in one-dimension so fluctuations depress the transition to the fully distorted Peierls' state and this factor must be taken into account in considering the behavior of these systems.[7]

A final factor which is unique to the one dimensional conductive systems is the effect of disorder.[8] In three dimensional systems weak disorder plays a rather minor role in transport properties, however, in one-dimensional systems even the smallest amount of disorder causes all the electron states to become completely localized. This presents a limiting factor for conductivity in the normal state but one could expect pair tunneling in a possible superconducting system to lead to bulk superconducting behavior. So this should present no fundamental limit to superconductivity.

In conclusion then we see that because organic systems usually form structures which are highly anisotropic or one-dimensional, they give rise to special problems when one considers the conductive or superconductive behavior. It should also be remarked that because the effective coherence length of the pair wavefunction can be expected to be large compared to the lattice constant one must organize the molecules in the structure over a considerable volume in order not to disrupt the superconducting state. This organization of the molecular preperties to yield a new state illustrates well the other meaning of "organic" to which Professor Akamatsu referred in his introductory remarks. It is here where I believe we have much to learn from studies of organic materials and which will bear rich fruit in the future.

REFERENCES

1. J. Bardeen, L. N. Cooper, and J. R. Schrieffer, Phys. Rev. 108, 1175 (1957).
2. W. A. Little, Phys. Rev. 134, A1416 (1964).
3. K. Krogmann, Angew. Chem. Int. Ed. 8, 35 (1969).
4. I. F. Shchegolev, Phys. Status. Solidi (a) 12, a (1972).
5. W. A. Little, Phys. Rev. 156, 396 (1967).
6. Y. M. Bychkov, L. P. Gor'kov, and I. E. Dzyaloshinskii, Soviet Physics, JETP 23, 489 (1966).
7. M. J. Rice and S. Strässler, to be published.
8. A. N. Bloch, R. B. Weisman, and C. M. Varma, Phys. Rev. Letters 28, 753 (1972).

Comments on the Metallic Conductivity in (TTF)(TCNQ) Complex

R. Aoki

Department of Physics Kyushu University

Fukuoka, 812 Japan

Relating to the subject of this seminar, an interesting topic has been reported[1,2] and discussed[3,4] by many groups, that is, large magnitude of metallic conductivity was measured in (TTF)(TCNQ)* charge transfer complex. Especially as high as $10^6 (\Omega\text{-cm})^{-1}$ a peak value was reported, as shown in Fig. 1, by a Penn. group [2] (L.B.Coleman, M.J.Cohen, D.J.Sandman, F.G.Yamagishi, A.F.Garito and A.J.Heeger, here after denoted by CCSYGH), and they ascribed this exceedingly large conductivity to one-dimensional superconducting fluctuations.

Among the points presented by CCSYGH as the evidence of super-conducting paraconductivity, however, some questions are found and I would like to discuss them.

i) CCSYGH say this large magnitude of conductivity σ to the extent of $10^6 (\Omega\text{-cm})^{-1}$ cannot be understood on the basis of conventional metallic conduction. In their estimation of anisotropic conductivity, a general expression

$$\sigma_x = \frac{2V}{2\pi} \cdot \frac{e^2 \tau}{\hbar} \cdot \int |v_{\kappa x}| \, dS_{Fx} \qquad (1)$$

is used[5], where x means easy conductive direction, in this case nearly perpendicular to the y-z molecular plane of (TTF) and (TCNQ), and dS_{Fx} is differential area perpendicular to x on the Fermi surface.

---

*   (TTF) is tetrathiofulvalene, and (TCNQ) is tetracyano-p-quinodimethane.

Today, there is little information on the Fermi surface of the carrier in (TTF)(TCNQ). So, CCSYGH introduced a crude assumption that the Fermi surface is limited to the 1st Brillouin zone as schematically shown in Fig.2(a).

It means one dimensional tight binding model and extremely large energy gap existence in y-z directions. But really most of TCNQ complexes have three dimensional conductivity even though σ_y and σ_z are less than 10^{-2} times of σ_x [1,6,7].

Similar anisotropic conduction is found in various organic compounds, and for the poorer conductive direction means less carrier mobility, larger effective mass and larger energy density of states, then the Fermi surface rather extends to the less conductive

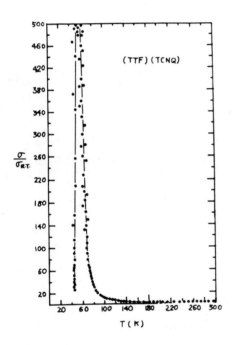

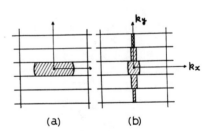

Fig. 1 Temperature dependence of the electrical conductivity σ of (TTF)(TCNQ) single crystal [from Penn. group[2]].

Fig. 2 Two models of the Fermi surface for (TTF)(TCNQ) complex; here its easy conductive direction corresponds to x-axis. (a) one-dimensional tight binding model; (b) a disk-like model extending in y-z plane.

directions. For example, graphite single crystal has been investigated and actually a sharp pencil like Fermi surface was found to
extend more than ten times in the less conductive c-direction([8,9]).

So it seems to be rather likely to presume the following
Fermi surface for (TTF)(TCNQ), that is a disk-like surface with axis
parallel to the easy conductive x-direction as shown in Fig.2(b).
The extension of the Fermi surface beyond the 1st Brillouin zone
in y-z directions may be accepted, considering some higher mode of
electronic structure within the unit cell of the crystal lattice([5]).
When this disk-like Fermi surface extends to the 3rd Brillouin
zone as shown in Fig.2(b), the cross section S_{FX} becomes about ten
times larger than that of the tight binding model in Fig.2(a).

If we use this large S_{FX} in eq. 1, the measured high conductivity σ, $10^6 (\Omega-cm)^{-1}$, can be understood with usual length
(i.e., several hundreds Å) of carrier mean free path, $l = v_F \cdot \tau$,
comparable to the case of copper, and without employing any
superconducting mechanisms.

ii) In temperatures above 58K in Fig. 1, the σ to T character is
found to show metallic conduction. CCSYGH plotted this σ-T relation
by $\log\sigma$ versus $\log(T-Tc)$ in Fig.3 and found a straight line
behavior expressed as

$$\sigma = \sigma_\bullet \cdot (T-Tc)^{-n} \qquad \text{with } n = 1.5 \text{ and } Tc = 58K.$$

They compared this result with theoretical expression based on
one-dimensional superconducting fluctuations where n=3/2 would be
expected, and proposed this straight line fitting with n=1.5 as a

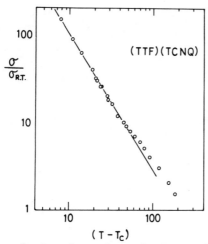

Fig. 3 Log-log plot of the conductivity
of (TTF)(TCNQ) single crystal vs. (T-Tc)
where Tc = 58K. The straight line has
slope of -1.5 [from Penn. group([2])]

powerful evidence for the existence of superconducting mechanism.

But I was some interested in, how regular non-superconducting metallic substances behave in the same $\log\sigma$ to $\log(T-Tc)$ scale plotting. Two simple conventional metals are chosen, one of which is a typical high conductive metal, copper, and the other, lead, whose Debye temperature Θ_D is[10] 108K comparable[11] with 90K of (TTF)(TCNQ).

In pure simple metals, electrical resistance mainly comes from thermal phonon scattering, and the resistivity R to temperature T relation is expressed by well known the Grüneisen-Bloch equation[12]. When the character is expressed as $R \sim T^{\alpha}$, and the exponent α is varing from 3 to 5 for $T \ll \theta_D$ and close to 1 for $\theta_D/5 < T$.

In Fig.4, we can see such a complicated T dependence of R is just reduced to remarkably simple a straight line relation in $\log\sigma$ to $\log(T-Tc)$ expression over one or two decades, by choosing an arbitrary parameter Tc properly. That is for copper Tc=40K and for lead Tc=10K. Each point was computed from the empirical data for bulk metal substances in Meaden's textbook[12], and the exponent n in the relation $\sigma = \sigma_o \cdot (T-Tc)^{-n}$ was found to be n=1.1±0.1 for both copper and lead.

The larger value of the exponent ($n \sim 1.5$) in (TTF)(TCNQ) than in the simple cubic metals ($n \sim 1.1$) may be understood as follows; the highly anisotropic phonon-field in (TTF)(TCNQ) probably leads to a non-degenerate phonon spectrum extending from low to high frequency regions; however, the Debye temperature (Θ_D=90K) determined from

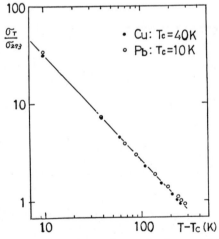

Fig. 4 Log-log plot of the conductivity σ of copper and lead bulk specimens vs. temperature (T-Tc) where each Tc was arbitraly chosen as denoted. The figure points were computed from the data in Meaden's textbook[12].

low temperature specific heats reflects[10] only the lower frequency
part in such a wide-spread phonon spectrum. So even in the relatively high temperature region ($\Theta_D/5 < T$), a non-linear relation in
R-T still may remain, that means n becomes larger than unity in
(TTF)(TCNQ), as contrasted with the case of simple cubic metals where
phonon field is degenerated and the Debye model approximates the
field well.

In any case, it turns out that the linear relation between $\log\sigma$
and $\log(T-Tc)$ cannot be used as a criterion for the existence of
superconducting paraconductivity, because obviously non-superconducting state bulk metals show similar behavior as shown in Fig.4.

iii) The BCS equation

$$Tc = \Theta_D \cdot \exp[-g^{-1}]$$

gives the relation between superconducting critical temperature Tc
and the Debye temperature Θ_D with electron-pair coupling parameter
g. Ginzburg and Kirzhnits predicted more extended utility of this
equation beyond the 3-dimensionality[13] and presented $g < 1/2$ as
the condition for realizing superconducting state[14]. In practice
to the present, no superconductors even in pseudo 2-dimensional
systems[15] are found outside of this rule.   Even if some
substances (e.g. A-15 crystal type compounds) would have relatively
large g values at high temperatures, they undergo martensitic
transformations in the course of cooling down and the g value is
well reduced to $< 1/2$ before going into superconducting state.

For (TTF)(TCNQ), Θ_D is given as 90K from low temperature specific heats[11], and the existence of a superconducting state was
required by CCSYGH with a critical temperature Tc supposed[2]
to be 58K.

Applying these values   to the BCS equation one gets $g \sim 2.5$ for
(TTF)(TCNQ). This value far exceeds the realizing condition $g < 1/2$,
hence it seems to be improbable to require the existence of superconductivity in (TTF)(TCNQ).

From the above mentioned view points i) ii) and iii), I would
comment that some reasonable understanding of (TTF)(TCNQ)'s
electrical conductivity characteristics both in large magnitude and
the temperature dependence in metallic region may be possibly
obtained by means of conventional conduction theory with more
careful considerations into its electronic and phonon states, without employing any other special mechanisms such as superconducting
paraconductivity. Further investigations about the Fermi surface
must, however, be done before practical models can be discussed.

REFERENCES

1) J.Ferraris, D.O.Cowan, V.Walatka and Jr.J.H.Perlstein: J.
   Amer. Chem. Soc. 95 (1973) 948.

2) L.B.Coleman, M.J.Cohen, D.J.Sandman, F.G.Yamagishi, A.F.Garito and A.J.Heeger: Solid State Comm. 12 (1973) 1125.
3) J.Bardeen: Solid State Comm. 13 (1973) 357.
4) P.Lee, T.M.Rice and P.W.Anderson: Phys. Rev. Lett. 31 (1973) 462.
5) A.J.Epstein, S.Etemad, A.F.Garito and A.J.Heeger: Phys. Rev. B5 (1972) 952.
6) W.J.Siemons, P.E.Bierstedt and R.G.Kepler: J. Chem. Phys. 39 (1963) 3523, 3528.
7) J.P.Farges: Phys. Letters 43A (1973) 161.
8) D.E.Soule, J.W.McClure and L.B.Smith: Phys. Rev. 134 (1964) A453.
9) A.P.Cracknell: Advance in Phys. 18 (1969) 756.
10) E.S.R.Gopal: Specific Heats at Low Temperatures (Plenum Press, New York, 1966).
11) T.Wei, S.Etemad, A.F.Garito and A.J.Heeger, to be published.
12) G.T.Meaden: Electrical Resistance of Metals (Plenum Press, New York, 1965).
13) V.L.Ginzburg and D.A.Kirzhnits: Soviet Phys. JETP 19 (1964) 269.
14) V.L.Ginzburg: Contemp. Phys. 9 (1968) 355.
15) F.J.Di Salvo, R.Schwall, T.H.Geballe, F.R.Gamble and J.H.Osiecki: Phys. Rev. Letters 27 (1971) 310.

DESIGN AND STUDY OF ONE-DIMENSIONAL ORGANIC CONDUCTORS I:

THE ROLE OF STRUCTURAL DISORDER

Aaron N. Bloch

Johns Hopkins University

Baltimore, Maryland, USA   21218

Much of this conference has been devoted to transport in materials which, like most organic solids, are relatively poor conductors of electricity. In this pair of talks I should like to address the rather different set of problems surrounding the small but growing class of organic compounds whose conductivities are quite substantial, even approaching those of common metals.

The current resurgence of interest in this topic is of course directed at the newest members of the class, the organic semimetals such as TTF-TCNQ,[1] and arises from the assertion[2] that a few samples of this material exhibit superconducting fluctuations near 60K. At Johns Hopkins, where these compounds were first studied,[1] we remain highly skeptical of these claims. We shall be discussing the reasons for this skepticism as part of the second paper, but this is not the central point of these talks. Rather, our thrust will be that if the controversy is to be resolved and TTF-TCNQ understood, it must be in the context of the general class of organic conductors of which this material is a member. We shall find that when the properties of the class are surveyed, there emerges a set of systematic chemical and physical criteria for the construction of an organic metal, and that from these criteria the synthesis of TTF-TCNQ follows rather naturally. From this perspective, we shall examine our present experimental and theoretical understanding of TTF-TCNQ and briefly discuss the prospects for further refinement in the design of such materials.

In general, one expects that intrinsic conductivity in purely organic solids will be limited by the large covalent energy gaps which stabilize the material. During the past decade, however, it has been shown that these difficulties can be circumvented in

certain special structures. Specifically, the Dupont group[3] was
the first to demonstrate that planar molecular ions having inhomo-
geneous charge densities can be stacked along the direction of
strong orbital overlap to form an array of parallel conducting
linear chains, well separated by channels containing the counter-
ions. The most prominent examples are salts of the organic electron acceptor TCNQ[3,4] (7,7,8,8-tetracyano-p-quinodimethane; see
Figure 1). The anisotropic ionic gap which stabilizes these
structures presents no impediment to one-dimensional band conduction along the chains, and the "quasi-one-dimensional" electronic
properties of the materials have been extensively documented.[4]

Where the stoichiometry of these compounds is such that the
one-dimensional conduction band is partially filled, we have the
intriguing prospect of metallic conductivity in an organic system.
It is well known, however, that conduction in narrow bands and
restricted geometries is subject to severe limitations. Indeed,
most TCNQ are not metals but magnetic insulators[4] displaying peri-
odically distorted chain structures[5] consistent with the instability of the one-dimensional electron gas[6] toward a charge[6] or spin-[7]
density wave of period π/k_F which opens a gap at the Fermi level.
In light of these observations, serious difficulties in interpretation were presented by the early discovery[3] that a very few TCNQ
salts do possess uniform chain structures and high conductivity at
room temperature. These exceptions include the simple salt of N-
methyl phenazinium (NMP) and the complex salts of quinolinium and
acridinium; the molecular structures appear in Figure 1. These
materials were the best organic conductors known prior to the synthesis of TTF-TCNQ, and represented a chemical puzzle: what special properties of the donor molecules of Figure 1 render their
TCNQ salts good conductors despite the instability of the one-dimensional electron gas, and despite the insulating behavior of
dozens of similar compounds?

The first systematic approach to this problem was the sugges-
tion by Le Blanc[8] that cation polarizabilities play an important
role in the conduction process. Where the cation is small and the
polarizability large, the attractive conduction electron- Frenkel
exciton interaction[9] can serve to enhance the conductivity by lowering the coulomb barrier to double occupation of a single TCNQ
site. There is force in this proposal: it is certainly true that
the insulating gaps are greatest in the TCNQ salts of non-polariz-
able donors,[3,4] and that the polarizabilities of all donors which
form highly conducting TCNQ salts are large. But it is also true
that there exist many equally polarizable electron donors whose
TCNQ salts are insulators,[3,4] and that an excessively large electron-
exciton interaction is itself expected to induce covalent instabilities.[10,11] Hence the widespread synthetic efforts to increase conductivities through exceptionally high cation polarizabilities
alone have been largely unsuccessful. It is necessary to look

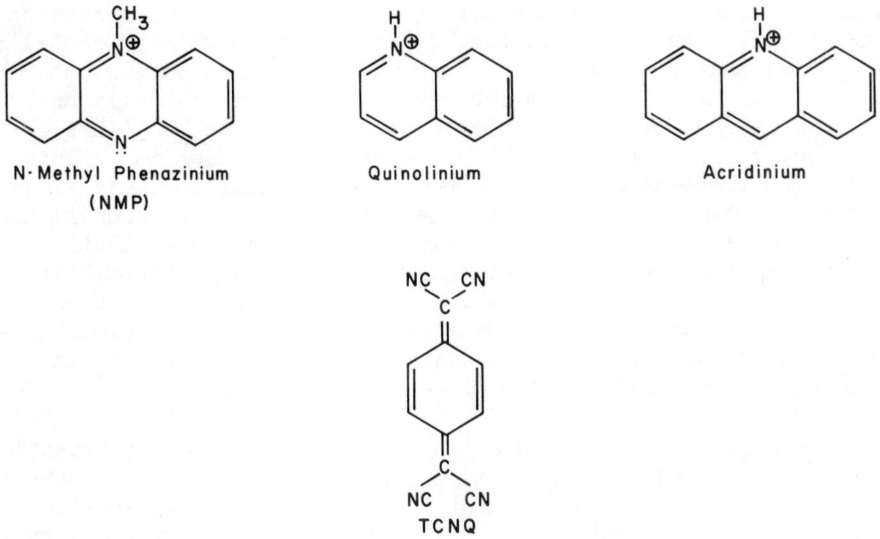

Figure 1 – Molecular structures of NMP, quinolinium, acridinium, and TCNQ.

further for the chemical prerequisites to high conductivity in these materials: the Le Blanc criterion is a necessary but not a sufficient condition.

A more telling factor was identified by Bloch, Weisman, and Varma.[12] We observed that in contrast with the symmetric constituent donor molecules of the distorted magnetic insulators, [3,4] the cations in Figure 1 possess permanent dipole moments typically larger than one Debye unit.[13] Further, x-ray crystallographic studies [3] show the dipoles to be distributed randomly between two inequivalent orientations. We reasoned that if Le Blanc[8] is correct and the interaction between conduction electrons and the relatively weak moments induced on the polarizable cations is large enough to reduce the gap substantially, the interaction between these electrons and the strong permanent moments must also be large enough to play an important role. For a carrier propagating along a given TCNQ chain, this interaction takes the form of a static potential which varies randomly from site to site. Its significance becomes apparent when one recalls the exact result[14] that in a one-dimensional system, all eigenstates of a random potential are localized.

We infer, then, that in these materials the partially occupied conduction band consists of a dense set of localized states. The form of the wavefunctions should reflect the competition between

localization due to the random potential, and coherent diffraction
by the otherwise periodic lattice. This situation we represented[12]
by supposing that along a chain, each wavefunction decays exponentially[14] at a rate α far from its center, but is essentially un-
attenuated over an extended central region of length L. (This
assumption, which we originally introduced [12] in order to explain
the detailed temperature dependence of the conductivity, has since
been verified by direct numerical claculation of the wavefunctions.[15])
As discussed below we deduce from experiment that L~5-20 lattice
constants, so that each state strongly overlaps many others. The
result is a class of materials which bear but limited resemblance
either to three-dimensional disordered systems or to one-dimensional metals. Rather, they universally display a unique set of electronic properties, which we now proceed to examine.

1. Structure An immediate consequence of the random potential
is a tendency toward a uniform chain structure.[12] The stability of
a one-dimensional distorted insulator arises from the square-root
singularity in the density of occupied states just below the insulating gap. The static potential fluctuations smooth the singularity
and place localized states in the nominal gap; for the rms fluctuation sufficiently large compared with the gap parameter, the gap is
effectively "washed out" and the uniform structure favored by
Madelung forces is recovered. This appears to be the situation in
the disordered one-dimensional organic conductors.

An interesting intermediate case is presented by the mixed-
valence platinum chain compounds,[16] where despite the presence of
structural disorder,[17] the soft phonon mode at $q=2k_F$ associated
with the Peierls instability[6] has been directly observed.[18] It is
difficult, however, to attribute the electrical and magnetic behavior to the Peierls transition alone. For example, the low-temperature magnetic susceptibility,[19] with its strong Curie tail, is not
that of a simple Peierls insulator; and the microwave dielectric
constant [4] of ca. 10^4 near 100K is more than an order of magnitude
larger than the optically determined contributions from the Peierls
soft mode[20] or direct interband transitions.[21] As discussed below,
this is precisely the behavior to be expected from a high density of
localized states at the Fermi level. It appears that the disorder
in these materials, though not strong enough to suppress the instability entirely, is sufficient to render the Peierls gap a pseudo-
gap.

Such a possibility is demonstrated by the recent numerical
calculation of Sen and Varma[22] on the Peierls transition in the
particular case of a substitutionally disordered binary alloy.
More general and physically more transparent, if less rigorous, is
the simple device of taking the Fermi wavenumber k_F as it appears
in the static density-density response function $\chi(q,o)$ to be complex,
with imaginary part equal to the reciprocal of the average localization length. The result is a complex $\chi(q,o)$ in which the

logarithmic singularity at $q=2k_F$ has been severely attenuated.
Applying the usual linear response theory[23] to calculate the temperature-dependent phonon frequencies, we find the mean-field
transition temperature T_p reduced by the disorder, and the observed
T_p of several hundred degrees[18] consistent with electron localization over 10-20 lattice constants in the platinum chain compounds.[24]
Further, from the imaginary part of the calculated phonon frequencies we deduce[24] a coherence length for the soft mode of more than
100 lattice constants, in agreement with experiment.[18] This difference between the one-electron and soft-mode correlation lengths
is not unexpected: the latter depend not only upon the former,
but also upon the unperturbed phonon frequency, the strength of the
electron-phonon coupling and the temperature.

2. <u>D.C. Conductivity.</u> Since all electronic states are localized, conduction can occur only through phonon-assisted hopping.[25]
Elsewhere[12,26,27] we have analyzed the anisotropic conductivity
in detail; here we simply summarize the results for the component
along the chain axis.

At low temperatures, most of the local hopping probablilities
are exponentially activated, and the conductivity contains important contributions from hopping over large distances between
states whose energies lie close to the Fermi level.[25] As $T\rightarrow 0$, this
becomes true even for hopping between chains, and we obtain[12,26]
for the conductivity σ the well-known three-dimensional result[25]:
$$\sigma \propto \exp[-(T_3/T)^{1/4}], \quad T\rightarrow 0 \qquad (1)$$
where T_3 is a constant. As the temperature is increased, the conductivity (1) rises less rapidly than does the contribution from
subnetworks in which hopping is confined to a single chain, and a
transition from three-dimensional to one-dimensional behavior[26]
occurs at a temperature $T_{\parallel}$, typically near 25K. Above $T_{\parallel}$, the
temperature dependence of the conductivity must be considered with
care. For a single, infinite, one-dimensional chain, the current
is determined by some very large resistance which cannot be bypass-
ed,[28] and will in general vary as $\exp(-A/T)$. The materials under
consideration, however, are more accurately represented as a large
array of parallel chains of finite length. In that case the distribution of limiting resistances must be averaged over all chains,[27,29]
and the resulting conductivity is found to be:
$$\sigma \propto \exp[-(T_1/T)^{1/2}], \quad T_{\parallel}<T<T_H \qquad (2)$$
Here $kT_1 = \alpha\lambda_1/\rho_0$, with ρ_0 the linear density of states and λ_1, a
critical constant, and $T_H=T_1/4\alpha^2L^2$ is the temperature at which the
largest important hopping distance falls below L. For $T>T_H$, long-
range hopping is no longer a factor, and the activation energy is
just equal to the average separation between states whose central
envelopes overlap a given segment of length L. This energy is:
$$kT_D\equiv(2\rho_0L)^{-1}=T_1/2\alpha L\lambda_1 \qquad (3)$$
As $T\rightarrow T_D$, the dominant hopping probabilities are but weakly temperature dependent and lead to a diffusive conductivity resembling that
of a metal having a very short mean free path.[12,26] We emphasize

that this is simply the high-temperature limit of the activated hopping process: <u>there is no true metal-to-insulator transition</u>. Within the simplest one-phonon approximation, the high-temperature conductivity is[26]

$$\sigma_D = Ne^2 \rho_o \gamma_o a^2, \quad T_D << T << T_F \tag{4}$$

where a is the lattice constant along the chains, T_F is the Fermi temperature, and γ_o is of the order of a phonon frequency.

The d.c. conductivities of the disordered one-dimensional conductors[4,16] follow these predictions closely.[12,26] Particularly satisfying is the self-consistency of the interpretation as indicated by the values of λ_1 deduced from experiment through the relation $\lambda_1^2 = T_1 T_H / T_D^2$. For seven compounds, λ_1 is found[12,30] to be 4.35 ± 0.35. The theoretical value has since been shown[27] to be 4.0.

The self-consistency extends to the pre-exponential factor, $\sigma_o(T)$, for the conductivity (2). This function is difficult to calculate precisely, but by considering the maximum number of critical resistors which can be physically placed in series along a given length of chain, one can easily set a lower bound.[31] For $T = T_H$, we find that $\sigma_o(T_H)/\sigma_D > 2\alpha L \lambda_1^{-1}(L/a)^2$. For NMP-TCNQ, with $\alpha L \sim 10$, this is consistent with the experimental ratio of $\sim 10^4$ if $L/a < 50$.

3. <u>Microwave Properties</u>. Particularly damaging to any interpretation of these materials as conventional intrinsic semiconductors at low temperatures is the observation[4,32] of microwave conductivities four to six orders of magnitude larger than the d.c. for $T \sim 10K$. Since the microwave frequency of 10 GHz is essentially zero on the scale of the apparent d.c. activation energy[4] of ~ 0.1 eV, the excess a.c. conductivity cannot be associated with transitions across the band gap in a simple semiconductor: some other low-frequency process is involved. This process also leads to an enormous positive microwave dielectric constant, $\epsilon(\omega) \sim 10^3 - 10^4$, at low temperatures.[4,32]

Such a process must certainly occur if the states at the Fermi level are localized. Associated with phonon-assisted hopping between these states are electronic Debye losses[25] qualitatively similar to those which arise from hopping between ionized impurities in a doped and compensated semiconductor. The present problem is distinct, however, in that for strongly overlapping localized states the independent-pair approximation of Pollak and Geballe[33] does not apply, and that the effect is of course much larger. Indeed, one finds that while the small excess microwave conductivity in the ordered, distorted insulator KTCNQ[34] can be attributed to $\sim 0.4\%$ charged impurities, to explain the effect in NMP-TCNQ on the same basis would require one "impurity" per lattice site.

Phenomenologically, we can describe the losses by considering the time-dependent response of a disordered one-dimensional conduc-

tor to an electric field $E(t)=E_o\theta(t)$, polarized along the chain
axis. Our concern here is not with the oscillatory terms in the
current corresponding to direct dipole transitions between strongly
overlapping localized states; these lead to an absorption proportional to ω^2 at low frequencies[25] but predominating in the infrared,
where it attains broad maximum.[21,35] Rather, we consider indirect,
phonon-assisted transitions excited by the applied field between
states not necessarily connected to the critical d.c. resistance
network, leading to a net polarization. The contribution of each
pair of states i and j to this polarization may be written

$$P_{ij}(t) = \alpha_{ij}[1-\exp(-\gamma_{ij}t)]E_o(t) \qquad (5)$$

where α_{ij} is a local polarizability which depends on the distance and
energy difference between states i and j. The polarization current
is given by the time derivative of (5), summed over all i and j;
its Fourier transform leads to the Debye equations averaged over hopping rates γ_{ij}. Explicit evaluation[36] yields a microwave conductivity which dominates the contribution from the critical resistance
network at low temperatures but approaches it at high temperatures,
in close agreement with experiment.

4. _Magnetic Susceptibility_. When a second electron is added to a
singly occupied localized state, the cost is some repulsion energy
U (much smaller, in general, than the intramolecular parameter U_o
[25,37] appearing in the Hubbard hamiltonian). It follows that at
T=0 a fraction of states at the top of the Fermi distribution will
have unpaired spins. After evaluating the grand partition function
for such a system with neglect of interactions between spins in
different localized states, we obtain for the temperature-dependent
magnetic susceptibility:

$$\chi(T) = \mu_B^2\rho_o(1-e^{-\beta U})^{-1/2} \ln\left|\frac{1+(1-e^{-\beta U})^{1/2}}{1-(1-e^{-\beta U})^{1/2}}\right| \qquad (6)$$

where μ_B is the Bohr magneton and $\beta\equiv(kT)^{-1}$. At high temperatures
such that $\beta U \to 0$ but $\beta E_F >> 0$, (6) becomes a Pauli susceptibility, but
at low temperatures $\beta U \to \infty$, it approaches the sum of a weaker Pauli
term and a Curie contribution from the singly occupied states.[12,25]
Equation (6) amounts to a specialization of the results of Kaplan,
Mahanti, and Hartmann,[38] and faithfully represents the experimental
behavior[4,37] of most disordered one-dimensional conductors. The
agreement is improved when the energy dependences of U and ρ_o are
included. For the half-filled-band material NMP-TCNQ,[37] however, a
tendency toward magnetic ordering at low temperatures cannot be
ignored. Crudely, this effect can be treated by approximating (6)
by its low-temperature limit, and including a small Weiss temperature θ. This expression has been fit to the data by Dr. V.V.
Walatka,[39] with the excellent result shown in Figure 2. From the
deduced value of ρ_o and the results of Sec. 2, we find that for
this material L/a<5. Support for these conclusions are drawn from
the recent work of Klein et al.,[40] who have shown theoretically
and experimentally that the susceptibility of NMP-TCNQ is not represented, as once supposed,[37] by a one-dimensional Hubbard model,

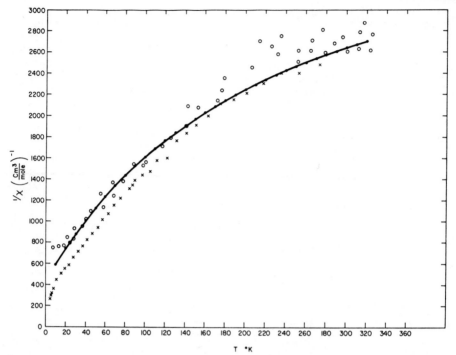

Figure 2. Magnetic Susceptibility of NMP-TCNQ. o-data of Epstein
et al.(Ref 38); x-data of Schegolev et al. (cf. Ref. 4); — - present
theory, with ρ_o=1.6 x 10^8 eV^{-1} cm^{-1} $\overline{U/k}$=104K, θ=23K.

but rather is dominated at low temperatures by localized moments.

5. Thermoelectric Power. The thermoelectric power for a
Fermi distribution of localized states is given by[25,41]
$$S(T)=(\pi^2k^2T/3e)\,(\partial\ln\sigma/\partial E)_{E_F} \qquad (7)$$
as for a metal. When (7) is evaluated using the results of Section
2, one finds that S(T) is linear in T, as though for a metal, when
T>T_H, and contains terms proportional to T and to $T^{1/2}$ at lower
temperatures. The relative importance of these terms depends on
the signs and magnitudes of the energy derivatives of α, L, and ρ_o
at the Fermi level. Within reasonable assumptions the experimental
data[4] are accurately described by (7)[36,39] except possibly at the
lowest temperatures, where the calculation is complicated by phonon
drag effects.

References

This and the following paper share a common set of references,
which are listed at the end of that paper.

DESIGN AND STUDY OF ONE-DIMENSIONAL ORGANIC CONDUCTORS II:

TTF - TCNQ AND OTHER ORGANIC SEMIMETALS

A.N. Bloch, D.O. Cowan, and T.O. Poehler

Johns Hopkins University

Baltimore, Maryland, USA 21218

What are the implications of the disordered one-dimensional conductors, discussed in the previous paper, for the development of organic materials of still higher conductivity? Simply, the situation is this. In lower-dimensional, narrow-band organic charge-transfer salts, we have systems in which the metallic state is inherently unstable. One way of suppressing the instabilities is by introducing sufficient disorder to "wash them out". As demonstrated by examples[12] such as NMP-TCNQ, this can lead in favorable circumstances to a spectacular increase in conductivity. Nevertheless the structural disorder remains itself a limiting factor. The localization of the electronic states at the Fermi level limits the conductivity to the diffusive range[25] of a few hundred reciprocal ohm-centimeters--enormous by organic standards, but low compared with good inorganic conductors. To elevate organic conductivities into the truly metallic range requires some way of suppressing the tendency toward a distorted insulator without resorting to the introduction of structural disorder. We therefore seek materials consisting of symmetric molecules with no strong dipole moment, and forming structures which offer more flexibility for design than does a single conducting chain.

Particularly simple and attractive alternatives are systems in which the donor and acceptor molecules each form quasi-one-dimensional chains, and in which populations of conduction electrons reside on both the donor chains ...DDDD... and the acceptor chains ... AAAA ... . We note that the stability of such a material is enhanced if more than one electron per donor molecule is available for transfer, so that the Madelung energy is large. Our attention is thereby focused on cases where both the donor and acceptor molecules can be divalent. We now have a structure in which there are an even number of conduction electrons per unit cell, so that metallic

167

conduction cannot occur unless two bands overlap in energy to form
a semimetal. Let us restrict our discussion to simple salts of
stoichiometry DA, and assume that neither D nor A has a low-lying
excited state within a typical bandwidth (say, a few tenths of an
electron volt) of the Fermi level. Then metallic behavior requires
that the conduction band which in a tight-binding representation
consists largely of donor states overlaps the one consisting largely
of acceptor states. The situation is summarized in Figure 1, where
we sketch schematically the projections of a simple one-electron
tight-binding band structure in directions parallel and perpendicular to the chain axis. We have taken the donor band D to be somewhat wider than the acceptor band A (as is likely for TTF-TCNQ),
but have assumed for simplicity a single parallel lattice constant
a and perpendicular lattice constant b. We ignore the differences

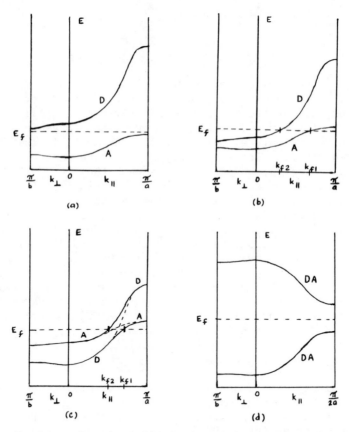

Figure 1. Band overlap in a divalent quasi-one-dimensional charge-
transfer salt DA. (a) Ionic insulator $D^{2+}A^{2-}$, with no band overlap. (b) Semimetal with small band overlap. (c) Semimetal with
large band overlap; bands split by covalent D-A interaction. (d)
"Covalent" insulator ...DADADA... .

between the first and second ionization potentials of D, and between
the first and second electron affinities of A; this is of course
chemically incorrect but has little effect on the simple qualitative
physics we discuss here.

Figure 1a depicts the situation where the difference between
the energies E_D^o and E_A^o of the free donor and acceptor ions is so
large that the two bands do not overlap. In that case the Fermi
energy E_F lies in the gap and we have an ionic insulator consisting
of D^{2+} and A^{2-} ions. Recent ESCA studies[42] suggest that this is
the case in the 1:1 salt of tetramethyltetrathiofulvalene and 1,2-
dichloro-4,5-dicyanoquinone (TMTTF-DDQ).

As $E_D^o - E_A^o$ is decreased, the bands eventually overlap, and the
material becomes a "one-dimensional" semimetal, as in Figure 1b.
From the Le Blanc[8] point of view (discussed in the previous paper),
the advantages of such a system are several. Since the bands are
not in general half-filled, intramolecular correlation effects
present a reduced impediment to metallic conduction.[8] More important, as Professor Little has observed at this conference, the
attractive interaction between a carrier and a Frenkel exciton[8,9]
is complemented by the much stronger attractive interaction between
an electron on one type of chain and a hole on the other. Since
the former interaction alone is found to reduce the effective intramolecular coulomb repulsion to ~ 0.1 eV in favorable cases,[8,37] one
has a hope that the latter can serve to diminish it further, to the
point where a metallic state is stable at ordinary temperatures.
As in the single-chain case, however, an excessively large attractive
interaction can itself induce an instability; for the band structure
of Figure 1b this takes the form of a dimerization of both chains
to form a one-dimensional excitonic insulator.[43] Fortunately one
finds, as we shall see, that by chemically adjusting the degree of
charge transfer from D to A, it is possible to optimize the strength
of the interchain electron-hole interaction at a level suitable for
metallic conduction.

This is not to imply that such a material is expected to remain
metallic at low temperatures. The two Fermi wavenumbers of Figure
1b are subject to the chemical constraint $k_{F1} + k_{F2} = \pi/a$, and the
band structure is unstable with respect to a Peierls distortion[6]
of wavenumber $q_\parallel = 2k_{F1} = g - 2k_{F2}$, where g is a reciprocal lattice vector.
Whether this takes place at a single transition temperature (i.e.,
$q_\perp = 0$) or at a separate temperature for each chain ($q_\perp = \pi/b$) depends
upon the details of the electron-phonon coupling, the interchain
coulomb interaction, and any temperature dependence of the degree
of charge transfer. In any event the transition temperature may be
somewhat reduced when k_{F1} and k_{F2} are not commensurate with the
lattice.

It is tempting to follow such reasoning further and reduce
$E_D^o - E_A^o$ still more, as in Figure 1c. Here the D and A bands actually

cross in reciprocal space, and are split by the interchain charge-transfer matrix element t_{DA}. If the crossing point occurs near the Fermi level, as in the figure, the one-dimensional constraint may be relaxed insofar as carriers are delocalized over both kinds of chain. It is important to recognize, however, that as we reduce $E_D^o - E_A^o$ we reduce the amount of charge transfer and hence the Madelung energy, destabilizing the entire structure. Ultimately, we may reach the point where it profits the system to rearrange in such a way as to maximize the splitting $2t_{DA}$- i.e., to arrange the donor and acceptor molecules face-to-face in ...DADADA... stacks.[5] This is the case, for example in the N,N,N',N'-tetramethyl-p-phenylene-diamine (TMPD)[5] salt of TCNQ. This instability represents a transition to a "covalent" semiconductor, as in Figure 1d. We conclude that the semimetallic state occurs only when $E_D^o - E_A^o$ is neither too large nor too small.

On the basis of these considerations we are able to enunciate a fairly coherent set of chemical criteria for the construction of an organic semimetal. We expect to achieve metallic conductivity in organic charge-transfer salts composed of linear chains of separately stacked donor and acceptor molecules. These should be:

    1. planar, with extensive π-electron systems and inhomogeneous charge distributions to facilitate stacking;

    2. comparatively small and polarizable, to satisfy the Le Blanc[8] criterion;

    3. symmetric, so as not to introduce intrinsic structural disorder;[12] and

    4. nominally divalent.

Given these characteristics, the important variables to be controlled in organic synthesis are (aside from steric factors) the first and second ionization potentials of the donor and electron affinities of the acceptor, and through them the intramolecular correlation energies, the bandwidths, the degree of band overlap and charge transfer, and the interchain coupling. The object is to manipulate these factors through the use of heteroatoms and substituent groups in order to achieve the proper balance for metallic conduction.

The first experimental realization of these criteria was the synthesis[1] of TTF-TCNQ, the best organic conductor known[1,2] and the first TCNQ salt of a symmetric cation (Figure 2) to exhibit a uniform chain structure[44] at room temperature. Both TTF (tetra-thiofulvalene) and TCNQ satisfy the desiderata (1)-(4) above, and each is known to form one-dimensional conducting salts of its own.[4,45] The crystal structure[44] of the 1:1 salt consists of parallel, separately stacked donor and acceptor chains (Figure 3), and we identify the material as the first organic semimetal.

Supporting this conclusion is the striking contrast between the electronic properties of TTF-TCNQ and those of the disordered

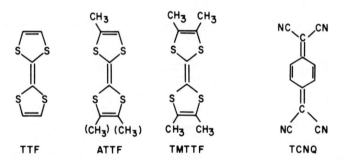

Figure 2.   Molecular structures of TTF, ATTF, TMTTF, and TCNQ

one-dimensional conductors. For TTF-TCNQ the maximum d.c. conductivity[1] of ~10^4 Ω^{-1} cm^{-1} near 60-70K is clearly in the metallic
range, nearly two orders of magnitude larger than the diffusive
limit[12] for the disordered materials. Below 60K the conductivity
is activated,[1] but cannot be fit to a law of the form $\ln\sigma \propto T^{-1/n}$.
The microwave conductivity[46] closely follows the d.c. at all temperatures, and the microwave dielectric constant[46] emphatically
reveals a metal-to-insulator transition near 60-70K. The contribution of the conduction electrons to the static magnetic
susceptibility[47] has no intrinsic Curie tail, but instead is near
zero below 60K and thermally activated above.

It was for this material that Coleman et al[2] reported the
observation of anomalously large conductivities in a very small
percentage of samples near 60K. We have several reasons to doubt
the validity of these experiments. First, despite widespread
intensive effort, the result has not been authenticated in any
other laboratory in the year since the preprint of Ref. 2 first
appeared. Second, measurements of the microwave conductivity,[46]
which are free of contact problems and less sensitive to gross
crystal defects, have been performed on well over 100 crystals with

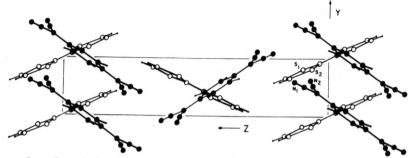

Figure 3.   Crystal structure of TTF-TCNQ.   o— TTF; • — TCNQ.

no sign of any such anomaly. Third, much of the case[2] for accepting non-reproducible giant conductivities in TTF-TCNQ rested upon
reproducible results for the (cis, trans) -dimethyl derivative
ATTF-TCNQ (Figure 2). Here because of sample problems Coleman et al[2]
were unable to measure the absolute magnitude of the conductivity,
but cited instead its strong temperature dependence for a silver-
paint compaction. We have since succeeded in measuring[46] the absolute microwave conductivity of single crystals of ATTF-TCNQ, and
found it to be comparatively small, with the reduced transition
temperature and partially "washed out" transition expected for a
weakly disordered material. Fourth, it is well known that inhomog-
eneous current distributions in highly anisotropic conductors can
lead to spuriously large apparent d.c. conductivities as measured
by the standard four-probe technique. Such an artifact has been
observed by Zeller[48] in the quasi-one-dimensional mixed-valence
platinum salts. More recently Shafer et al[49] have shown experimentally that spurious anomalies similar to the results of Coleman
et al[2] can be induced in TTF-TCNQ by a simple misalignment of contacts.

Finally, if one does assume that the results of Ref. 2 are
valid, the most plausible theoretical explanation is the specula-
tion[50,51] that conduction above 60K is dominated by Frohlich's[52]
travelling charge-density wave. In that case the maximum in the
conductivity near 60K would be associated[51] with a pinning of the
current-carrying collective mode rather than with the opening of a
gap in the single-particle spectrum. Yet the far-infrared photoconductivity[53] at 4K shows just such a gap, of magnitude corresponding precisely to a transition temperature of 70K.

Inasmuch as the temperature dependence of the magnetic
susceptibility[47] implies a much higher mean-field transition temperature,[47,51] this last observation may be interpreted[53] as weak
evidence that the TTF and TCNQ chains undergo separate Peierls
distortions at quite different temperatures. In this context it
is interesting that the tetramethyl analog TMTTF-TCNQ[54] (Figure 2),
which has a similar structure[55] and temperature-dependent
susceptibility,[56] shows in the microwave a metal-to-insulator
transition near room temperature suggesting that the two chains
are no longer independent. This is consistent with the stronger
interchain coulomb coupling expected to accompany the increased
charge transfer which results from the lower ionization potential
of the TMTTF donor.

The series (TTF, ATTF, TMTTF)-TCNQ is a very simple example
of the development of rational chemical trends in organic conductors. We believe that our physical understanding of these materials is now reaching the point where systematic chemical control of
their electronic properties is possible.

References (Papers I and II)

1. J.P. Ferraris, D.O. Cowan, V. Walatka, and J.H. Perlstein, J. Am. Chem. Soc. $\underline{95}$, 948 (1973).
2. L.B. Coleman et al, Sol. St. Comm. $\underline{12}$, 1125 (1973).
3. W.J. Siemons et al, J. Chem. Phys. $\underline{39}$, 3523 (1963); R.G. Kepler, J. Chem. Phys. $\underline{39}$, 3528 (1963); L.R. Melby, Can. J. Chem. $\underline{43}$, 1448 (1965).
4. For a review, see I.F. Schegolev, Phys. Stat. Solidi $\underline{A12}$, 9 (1972).
5. F.H. Herbstein, in J.D. Dunitz and J.A. Ibers, eds., Perspectives in Structural Chemistry, Vol. IV (1971).
6. R.E. Peierls, Quantum Theory of Solids (University Press, London, 1955), p. 108.
7. A.W. Overhauser, Phys. Rev. Lett. $\underline{4}$, 462 (1960).
8. O.H. Le Blanc, J. Chem. Phys. $\underline{42}$, 4307 (1965).
9. W.A. Little, Phys. Rev. $\underline{A134}$, 1416 (1964).
10. R.E. Merrifield, J. Chem. Phys. $\underline{44}$, 4005 (1965).
11. J.C. Phillips, Phys Rev. Lett. $\underline{29}$, 1551 (1972).
12. A.N. Bloch, R.B. Weisman, and C.M. Varma, Phys. Rev. Lett. $\underline{28}$, 753 (1972).
13. According to CNDO results kindly supplied by R. Metzger (private communication).
14. R.E. Borland, Proc. Roy. Soc. (London) $\underline{A274}$, 529 (1963).
15. C. Papatriantafillou, Phys. Rev. $\underline{B7}$, 5386 (1973).
16. H.R. Zeller, Feskorperprobleme XIII, 31 (1973).
17. K. Krogmann, Angew. Chem. Int. Ed. $\underline{8}$, 35 (1969).
18. B. Renker et al, Phys. Rev. Lett. $\underline{30}$, 1144 (1973).
19. A. Menth and M.J. Rice, Sol. St. Comm. $\underline{11}$, 1025 (1973).
20. P. Bruesch, M.J. Rice, S. Strassler, and H.R. Zeller (preprint).
21. J. Bernasconi, P. Bruesch, D. Kuse and H.R. Zeller, J. Phys. Chem. Solids (in press).
22. P. Sen and C.M. Varma, Bull. Am. Phys. Soc. $\underline{19}$, 49(1974).
23. e.g., M.J. Rice and S. Strassler, Sol. St. Comm. $\underline{13}$, 125 (1973).
24. A.N. Bloch and B. Huberman, unpublished work.
25. e.g., N.F. Mott and E.A. Davis, Electronic Processes in Non-Crystalline Materials (Oxford University Press, Oxford, 1971).
26. A.N. Bloch and C.M. Varma, J. Physics C $\underline{6}$, 1849 (1973).
27. V.K.S. Shante, C.M. Varma, and A.N. Bloch, Phys. Rev. $\underline{B8}$, 4885 (1973).
28. J. Kurkijarvi, Phys. Rev. $\underline{B8}$, 922 (1973).
29. W. Brenig, G. Dohler, and H. Heyzenau , Phil Mag, $\underline{27}$, 1093 (1973).
30. T.W. Thomas et al, J. Chem. Soc. $\underline{1972}$, 2050.
31. The oversimplified diffusive model of Ehrenfreund et al [Phys Rev. Lett. $\underline{29}$, 269 (1972)], when purged of errors, satisfies the inequality and gives $\sigma_0/\sigma_D \sim 4\sqrt{2} \ \alpha L \ (L/a)^2$.

32. T.O. Poehler, unpublished data.
33. M. Pollak and T.H. Geballe, Phys. Rev. Lett. 122, 1742 (1961).
34. R.M. Vlasova et al, Sov. Phys. Sol. St. 12, 2979 (1971).
35. Y. Iida, Bull. Chem. Soc. Japan 42, 71, 637 (1969).
36. A.N. Bloch, unpublished work.
37. A.J. Epstein et al, Phys. Rev. B5, 952 (1972).
38. T.A. Kaplan, S.D. Mahanti, and W.M. Hartmann, Phys. Rev. Lett 27, 1796 (1971).
39. V. Walatka, Ph. D. thesis, Johns Hopkins University (unpublished).
40. D.J. Klein, W.A. Seitz, M.A. Butler, and Z.G. Soos, preprint.
41. M. Cutler, Phil. Mag. 24, 173 (1971).
42. M.A. Butler, J.P. Ferraris, A.N. Bloch, and D.O. Cowan, Chem. Phys. Lett. (in press).
43. e.g., B.I. Halperin and T.M. Rice, Rev. Mod. Phys. 40. 755 (1968).
44. T.E. Phillips et al, Chem. Comm. 1973, 471; T.J. Kistenmacher, T.E. Phillips, and D.O. Cowan, Acta Cryst. (in press).
45. F. Wudl et al, J. Am. Chem. Soc. 94, 670 (1972).
46. A.N. Bloch, J.P. Ferraris, D.O. Cowan, and T.O. Pcehler, Sol. St. Comm. 13, 753 (1973).
47. J.H. Perlstein, J.P. Ferraris, V. Walatka, D.O. Cowan, and G.A. Candela, Proc. Conf. Mag. and Mag. Mats., Nov. 1972.
48. H.R. Zeller, private communication.
49. D.E. Shafer, G.A. Thomas, F. Wudl, J.P. Ferraris, and D.O. Cowan, Sol. St. Comm. (in press).
50. J. Bardeen, Sol. St. Comm. 13, 357 (1973).
51. P.W. Anderson, P.A. Lee, and M. Saitoh, Sol. St. Comm. 13, 595 (1973); P.A. Lee, T.M. Rice, and P.W. Anderson, Phys. Rev. Lett. 31, 462 (1973); Sol. St. Comm. (in press).
52. H. Fröhlich, Proc. Roy. Soc. A223, 296 (1954).
53. T.O. Poehler, A.N. Bloch, J.P. Ferraris, and D.O. Cowan, preprint.
54. J.P. Ferraris, T.O. Poehler, A.N. Bloch, and D.O. Cowan, Tet. Lett. 27, 2553 (1973).
55. T.J. Kistenmacher, private communication.
56. Y. Tomkeiwicz, private communication.

VII. SOLID MOLECULAR COMPLEXES

THE STRUCTURES AND PROPERTIES OF SOLID MOLECULAR

COMPLEXES AND RADICAL SALTS: A BRIEF REVIEW OF RECENT

STUDIES IN JAPAN

HARUO KURODA

DEPARTMENT OF CHEMISTRY, FACULTY OF SCIENCE

THE UNIVERSITY OF TOKYO

A considerable number of Japanease scientists are actively working on the problems related to the structures and properties of solid molecular complexes and radical salts, and it is impossible to take up all of their work in this very brief review. Thus I will restrict my talk to a few topics which seem interesting from the aspect of the study on the relation between the electronic behavior and crystal structure.

OPTICAL PROPERTIES OF TCNQ SALTS

On a variety of TCNQ salts, the diffuse reflection spectra of powders were examined by Y. Iida of Hokkaido University,[1] who pointed out a similarity between the spectrum of a TCNQ simple salt and the absorption spectrum of the TCNQ anion dimer formed in solution. The spectra of the complex salts of high electrical conductivity were considerably different from the simple salts, and characterized, in particular, by the presence of a low-energy absorption band extending to infrared region.

In order to elucidate the nature of the absorption bands, the polarized absorption spectra were observed in our laboratory on the single crystals of $K(TCNQ)$ and $Cs_2(TCNQ)_3$ by using the microspectrophotometeric method.[2] On the basis of the exciton theory, we estimated the factor-group splitting, crystal shift and crystal-induced mixing of the local-excitation bands, and gave interpretations for the observed crystal spectra.

From the analysis of the specular reflection of pressed-pellet, T. Sakata, Y. Oohashi and S. Nagakura in the Institute of Solid State Physics of the University of Tokyo tried to determine the absolute extinction coefficient and refractive index on several TCNQ salts.[3] The extinction coefficient data reported by them suggest the presence of considerable crystal-induced mixing among various excited states including charge-transfer states. It is of particular interest that the refractive index was found to be appreciably dependent on the wavelength, especially in the region of a strong charge-transfer band.

PHASE TRANSITIONS OF TCNQ SALTS

$[(C_6H_5)_3PCH_3] \cdot (TCNQ)_2$ is known to exhibit a phase transition at about 315 K, where it shows a discontinuous change of electrical conductivity as well as an appreciable change in the magnetic properties.[4] The heat capacity of this radical salt in 11-350 K region was measured by S. Seki and his collaborators in Osaka University.[5] According to their results, the entropy of transition is 1.7206 cal/deg.mol. When estimated from the magnetic and electrical data, the contributions of spins and charge carriers are expected only as 0.13 and 1.5×10^{-3} cal/deg. mol, respectively. Thus, Y. Iida attributed the main origin of the observed entropy of transition to the structural changes in the phase transition.[6]

The crystal structures of the low- and high-temperature phases of $[(C_6H_5)_3PCH_3] \cdot (TCNQ)_2$ were determined by Y. Saito and his collaborators in the Institute of Solid State Physics of the University of Tokyo.[7] In the case of a radical salt with one-dimensional column of radical ions, one can generally expect a phase transition associated with the Peierls distortion. In effect, this type or phase transition has been reported on several TCNQ salts. Interestingly, however, the arrangement of TCNQ anion radicals was found to be little different between the low- and high-temperature phases of $[(C_6H_5)_3PCH_3] \cdot (TCNQ)_2$. It is in the orientations of phenyl groups of $[(C_6H_5)_3PCH_3]^+$ ion that a significant change occures at the phase transition. This is rather unexpected and seems to offer an interesting problem to be discussed.

$[(C_6H_5)_3PCH_3] \cdot (TCNQ)_2$ and $[(C_6H_5)_3AsCH_3] \cdot (TCNQ)_2$ form a solid solution. By means of differential scanning calorimetry, Y. Iida studied the phase transition of $[(C_6H_5)_3PCH_3]_{1-x} \cdot [(C_6H_5)_3AsCH_3]_x \cdot (TCNQ)_2$ ($0 \leq x \leq 1$) system.[8] The heat capacity of the same system was studied by S. Seki and his collaborators.[9]

On the alkali metal-TCNQ simple salts, S. Minomura and

his collaborators in the Institute of Solid State Physics of the University of Tokyo studied the variation of electrical conductivity under a high pressure up to 15 k bar, and found a pressure-induced phase transition.[10]

OPTICAL PROPERTIES OF WÜRSTER'S SALTS

The polarized absorption spectra of single crystals of Würster's salts were observed at the room temperature by J. Tanaka and M. Mizuno in Nagoya University, who carried out a theoretical calculation on the crystal spectra to establish their interpretation.[11] The crystal structures of some of these salts were determined in the same laboratory.

As is well known, Würster's blue perchlorate exhibits a phase transition at 186 K, where the arrangement of Würster's blue cation radicals changes from monomeric type to dimeric one. T. Sakata and S. Nagakura observed the temperature dependence of the charge-transfer band in the absorption spectrum of powder, and analysed the data by assuming the singlet-triplet model.[12] The J-value thus determined from the spectroscopic data was 235 cm^{-1} in good agreement with the J-value, 246 cm^{-1}, reported from the ESR experiment.

We examined the temperature dependence of the polarized absorption spectrum of Würster's blue perchlorate single crystal.[13] The temperature dependence of the charge-transfer band in the single crystal spectrum was confirmed to be essentially same as that reported from the powder spectrum.

PHASE TRANSITIONS OF ALKALI METAL-CHLORANIL ANION RADICAL SALTS

The pressure effect on the powder spectra of alkali metal-chloranil anion radical salts was studied by S. Minomura and his collaborators.[14] According to their results, the intensity of the absorption band in the 10000-15000 cm^{-1} region, which was tentatively assingned by them to a charge-transfer band, markedly increases on application of high pressure.

The crystal structure of K·(chloranil) was determined by Y. Saito et al, who showed that chloranil anion radicals are stacked to form a column parallel to the a-axis.[15] In the polarized absorption spectrum of the single crystal observed in our laboratory, the low-energy band at 11500 cm^{-1} is completely polarized in the a-axis direction, indicating that this band is indeed a charge-transfer band associated with the interaction between chloranil anion radicals.[16] We examined the temperature dependence of the single crystal spectrum, and found that the intensity of the charge-transfer band markedly increases as the temperature is lowered below the transition point which has been found at 210 K from the study on the magnetic property. As known for the Würster's blue perchlorate, there is a close correlation between the intensity change of the charge-transfer band and the temperature dependence of the paramagnetic susceptibility.

We found further that K·(chloranil) salt has a tendency to take water molecules into the crystal lattice to be a hydrate. Some organic solvents, such as CH_2Cl_2 and CH_3CN, can also enter into the lattice of K·(chloranil) salt. [17)

SOLVENT-CONTAINING MODIFICATIONS OF BENZIDINE-TCNQ COMPLEX

In spite of the relatively low ionization potential of benzidine, the solid molecular complex composed of benzidine and TCNQ is not a radical salt, but a charge-transfer complex of non-ionic ground state. Interestingly, various solvent molecules can enter into the crystal lattice of benzidine-TCNQ to form a solvent-containing modification.[18) The examples of the solvent that gives a solvent-containing modification, are CH_2Cl_2, C_2H_5Cl, CH_3CN, CH_3COCH_3 and benzene. The electrical conductivity of a solvent-containing crystal is appreciably different from the solvent-free crystal, and varies depending on the kind of the solvent molecules contained in the crystal lattice. A considerable difference was found also in the absorption spectrum of the crystal.

We carried out the crystal structure analysis on these solvent-containing modifications of benzidine-TCNQ complex.[19) Both in the solvent-free and solvent-containing crystals, benzidine and TCNQ molecules are alternately stacked on each other to form a mixed molecular column like a typical charge-transfer complex. In the solvent-containing crystal, there are channels between the columns running parallel to them, in which solvent molecules are accomodated, whereas benzidine-TCNQ columns are closely packed in the solvent-free crystal. It was revealed that, in a solvent-containing crystal, hydrogen bonds are formed between the neighboring benzidine-TCNQ columns, which is likely to be playing an important role in the formation of the solvent-containing crystal.

SOME OTHER STUDIES

In a complex formed between aromatic amine and polynitro-phenol, there are two possible types of interaction between the components: the one is the formation of ions by the proton transfer, and the other is a charge-transfer interaction.

Y. Matsunaga and his collaborators in Hokkaido University are carring out a systematic investigation on such a system in order to elucidate the roles of the proton transfer and charge transfer on the complex formation.[20]

As already cited in this review, the crystal structure analysis has been carried out on various molecular complexes and radical salts in several laboratories, especially, by Y. Saito's group. Among these investigations, I wish to add a few other interesting examples. In phenothiazine-TCNQ complex,H. Kobayashi and Y. Saito found an unusual long-range sinusoidal structure of phenothiazine-TCNQ column, which seemingly due to hydrogen bonds between the neighboring column.[21] The brown and green forms of 1,6-diaminopyrene-chloranil complex, which were found eariler by Y. Matsunaga, were studied by T. Uchida et al, who showed that, although the brown-form crystal has a crystal structure typical to a charge-transfer complex, the overlap between the donor and acceptor is very little in the green-form crystal, where hydrogen bonds between 1,6-diaminopyrene and chloranil seem to play an important role for the complex formation.[22]

1) Y. Iida; Bull. Chem. Soc. Japan, 42, 71, 637(1969)
2) S. Hiroma, H. Kuroda, and H. Akamatu; Bull. Chem. Soc. Japan, 44, 9(1971)
3) T. Sakata, Y. Oohashi, and S. Nagakura; Molecular Structure Symposium of Chem. Soc. Japan (Kyoto, 1971)
4) a) Y. Iida, M. Kinoshita, M. Sano and H. Akamatu; Bull. Chem. Soc. Japan, 37, 428(1964)
   b) Y. Iida, M. Kinoshita, A. Kawamori and K. Suzuki; ibid., 37, 764(1964)
5) A. Kosaki, Y. Iida, M. Sorai, H. Suga, and S. Seki; Bull. Chem. Soc. Japan, 43, 2280(1970)
6) Y. Iida; Bull. Chem. Soc. Japan, 44, 3344(1971), 46, 320 (1973)
7) a) H. Tsuchiya, H. Kobayashi, F. Marumo and Y. Saito; Molecular Structure Symposium of Chem. Soc. Japan (Tokyo, 1970)
   b) M. Konno and Y. Saito; 28th Annual Meeting of Chem. Soc. Japan (1973)
8) Y. Iida; Bull. Chem. Soc. Japan, 43, 3685(1970)
9) A. Kosaki, M. Sorai, H. Suga and S. Seki; 26th Annual Meeting of Chem. Soc. Japan (1972)
10) N. Sakai, I. Shirotani and S. Minomura; Bull. Chem. Soc. Japan, 45, 3314, 3321(1972)
11) J. Tanaka and M. Mizuno; Bull. Chem. Soc. Japan, 42, 1841 (1969)
12) T. Sakata and S. Nagakura; Bull. Chem. Soc. Japan, 42,

1497(1969)
13) K. Kaneko and H. Kuroda; Molecular Structure Symposium of
Chem. Soc. Japan (Kyoto, 1971)
14) N. Sakai, I. Shirotani and S. Minomura;
Bull. Chem. Soc. Japan, 44, 675(1971)
15) M. Konno, H. Kobayashi, F. Marumo and Y. Saito;
Bull. Chem. Soc. Japan 46, 1987(1973)
16) S. Hiroma and H. Kuroda; Bull. Chem. Soc. Japan 46, (1973)
(to be published)
17) S. Hiroma and H. Kuroda; Molecular Structure Symposium of
Chem. Soc. Japan (Sendai, 1972)
18) M. Ohmasa, M. Kinoshita and H. Akamatu;
Bull. Chem. Soc. Japan, 44, 391(1971)
19) a) H. Kuroda, I. Ikemoto, K. Yakushi and K. Chikaishi;
Acta Cryst., S4, 5(1972)
b) I. Ikemoto, K. Chikaishi, K. Yakushi and H. Kuroda,
Acta Cryst. B28, 3502(1972)
c) K. Yakushi, I. Ikemoto and H. Kuroda; Acta Cryst. B29,
(1973) (to be published)
20) a) G. Saito and Y. Matsunaga; Bull. Chem. Soc. Japan, 44,
3328(1971)
b) Y. Matsunaga and G. Saito; ibid., 45, 963(1972)
c) N. Inoue and Y. Matsunaga; ibid., 45, 3478(1972)
d) G. Saito and Y. Matsunaga; ibid., 46, 714(1973)
e) Y. Matsunaga; ibid., 46, 998(1973)
21) H. Kobayashi and Y. Saito; Molecular Structure Symposium of
Chem. Soc. Japan (Kyoto, 1971)
22) a) T. Uchida, K. Kimura and K. Hoshino; Molecular
Structure Symposium of Chem. Soc. Japan (Kyoto, 1971)
b) T. Uchida, K. Kimura and T. Tokumoto;
26th Annual Meeting of Chem. Soc. Japan (1972)

MICROWAVE DIELECTRIC CONSTANTS OF SOLID MOLECULAR COMPLEXES

K. ISHII*, M. KINOSHITA* and H. KURODA

Department of Chemistry, Faculty of Science
the University of Tokyo, Hongo, Tokyo

Although the dielectric properties are of great significance for understanding various electronic processes such as SCL current, the dielectric constant has been reported only on a limitted number of organic crystals. This is primarily due to the experimental difficulties associated with the requirements given on the shape and the size of the specimen in the conventional method. We have tried to determine dielectric constant by the cavity perturbation method in the X-band microwave region.

The block diagram of our apparatus is shown in Fig. 1. The microwave generated by a klystron is introduced into the cavity. The klystron is frequency-swept around the resonance frequency. The resonance is detected with a crystal detector, and displayed on an oscilloscope.

The crystalline powder of the sample is packed in a quartz tube of an outer diameter of 0.40 cm, which is inserted into the cavity. The resonance frequency of the cavity can be related to the apparent dielectric constant of the specimen, ϵ_s, by the following equation.[1]

$$\frac{f_1^2 - f_0^2}{f_1^2} = -(\epsilon_s - 1)\frac{\int_{V_1} E_0 \cdot E_0^* \, dv}{\int_V E_0 \cdot E_0^* \, dv} + A(\epsilon_s - 1)^2,$$

---

* Present address: The Institute for Solid State Physics, the University of Tokyo.

where f_1 and f_0 are the resonance frequencies with and without the specimen respectively, V_1 and V are the volume of the specimen and that of the cavity, E_0 is the electric field vector in the absence of the specimen, and A is a factor for the second order effect. Knowing the volume fraction of the sample, δ, in the quartz tube, we calculated the dielectric constant of the crystal, ϵ, from ϵ_s by using Böttcher's equation,[2]

$$\epsilon = \frac{3\delta\epsilon_s + 2\epsilon_s(\epsilon_s - 1)}{3\delta\epsilon_s - (\epsilon_s - 1)}.$$

Although the value of ϵ thus obtained is slightly dependent on δ, we have found by examing the data of sample of known dielectric constant that the ϵ-value for $\delta = 0.5$ is best to be taken as the dielectric constant of the crystal.[1]

The dielectric constants of aromatic hydrocarbons are listed in Table 1, together with the molecular polarizations calculated from the observed dielectric constants. The molecular polarizations estimated by assuming the additivity of the bond polarizations, for which we adopted the values proposed by Le Fèvre[3] as "electronic" bond polarizations, are given in the third column of Table 1. These estimated values are in good agreement with the observed ones, the difference between the observed and estimated values being less than 2 cm^3/mol.

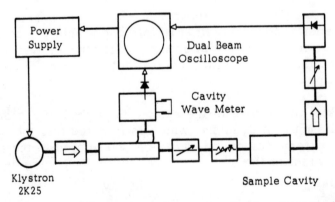

Figure 1.   Schematic Diagram of the Apparatus. The sample cavity; TE_{105} mode rectangular cavity for room temperature measurement, TM_{012} mode cylindrical cavity for variable temperature measurement.

Table 1.   Dielectric Constants and Molecular Polarizations
            of Aromatic Molecular Crystals

Substance	ϵ	P_{ob}*	P_{calc}*	Difference
Naphthalene	2.87	42.1	41.2	0.9
Anthracene	3.12	59.0	57.2	1.8
Phenanthrene	2.96	58.3	57.2	1.1
Pyrene	3.14	66.3	64.8	1.5
Chrysene	3.09	73.6	73.2	0.4
Perylene	3.34	81.9	80.8	1.1
Durene	2.55	44.4	43.3	1.1
Acenaphthene	2.94	49.8	48.2	1.6
Biphenyl	2.88	50.3	49.5	0.8
p-Terphenyl	2.98	74.4	73.9	0.5

* Unit; cm^3/mol.

Table 2.   Dielectric Constants and Molecular Polarizations
            of TCNQ Complexes

Donor	ϵ	P_{DA}*	$\Delta P = P_{DA} - (P_D + P_A)$*
Anthracene	3.30	132.3	3.9
Phenanthrene	3.4	129.7	2.0
Pyrene	3.42	140.1	4.4
Chrysene	3.55	147.4	3.4
Perylene	3.71	166.0	14.7
Phenothiazine**	5 - 6	165 - 180	33 - 54
Benzidine**	6 - 7	184 - 196	41 - 60
TMPD - TCNQ	5.4 - 5.9		
$K^+(TCNQ)^-$	5.4 - 5.9		

TCNQ;   3.48   69.4 $(: P_A)$

* Unit; cm^3/mol.
** The dielectric constants of phenothiazine and benzidine were
   estimated to be 2.9 - 3.2, 3.7 - 4.0, respectively.

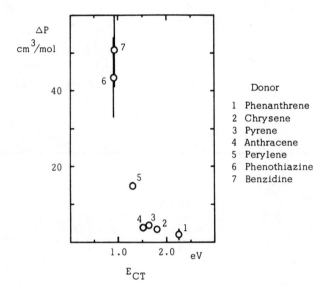

Fig. 2   ΔP plotted vs E_{CT}.

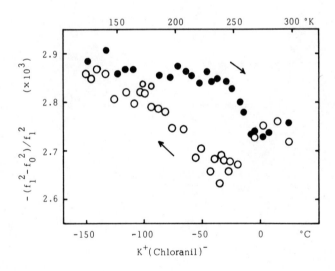

Fig. 3   Temperature Dependence of Frequency Change.

The data for solid molecular complexes which involve TCNQ as the electron acceptor, are given in Table 2. As are listed in the first column of this table, some of these complexes show dielectric constants larger than either donor substance or TCNQ. In the third column of this table, we have given the difference between the molecular polarization observed of the molecular complex, P_{DA}, and the sum of the observed polarization of the donor, P_D, and that of the acceptor, P_A. The difference, ΔP, is evidently larger than the experimental error. In the cases of phenothiazine-TCNQ and benzidine-TCNQ, the difference amounts to more than 30 cm^3/mol.

All these molecular complexes, except TMPD-TCNQ and K$^+$(TCNQ)$^-$, have a non-ionic ground state where the amount of charge-transfer is quite small. Thus we can expect that the contribution of lattice polarization is negligible, and the observed difference is primarily due to the increase in the electronic polarization on complex formation. As is well known, a molecular complex has charge-transfer excited states besides the locally-excited states associated with the excitations either in the donor molecule or in the acceptor molecule. The contribution of these charge-transfer states to the polarizability would be roughly proportional to $\Sigma f_k/E_k^2$, where f_k and E_k are respectively the oscillator strength and the excitation energy of the k-th charge-transfer state. Usually the lowest charge-transfer state gives the largest contribution. Thus one can expect that the polarizability will be larger with the decrease in the excitation energy and/or the increase in the oscillator strength of the lowest charge-transfer transition.

In Fig. 2, we have plotted ΔP against the energy of the lowest charge-transfer transition, E_{CT}. There is a clear tendency that ΔP increase on decreasing E_{CT}.

Other data which demonstrates the effect of the charge-transfer interaction is shown in Fig. 3, where the observed values of $-(f_1^2 - f_0^2)/f_1^2$ for K$^+$(chloranil)$^-$ are plotted against the temperature. This data shows that the dielectric constant of K$^+$(chloranil)$^-$ increases as the temperature is lowered below the transition point where the crystal structure is considered to change into the dimeric-form.[4] On this complex, the intensity of the charge-transfer band has been known to increase on lowering the temperature.[5] Thus there seems to be a parallel relation between the increase of the dielectric constant and the intensity of the charge-transfer band.

1) K. Ishii, M. Kinoshita and H. Kuroda, Bull. Chem. Soc. Japan, 46, 3385, (1973).

2) C.J.F. Böttcher, "Theory of Electric Polarization," Elsevier, Amsterdam, (1952).

3) R.J.W. Le Fèvre, "Advances in Physical Organic Chemistry," vol.3, ed. by V. Gold, Academic Press, London, (1965), p1.

4) J.J. Andre and G. Weill, Chem. Phys. Letters, 9, 27, (1971).

5) S. Hiroma and H. Kuroda, Bull. Chem. Soc. Japan, 46, (1973), (to be published).

ELECTRICAL AND OPTICAL PROPERTIES OF THE PHENOTHIAZINE-IODINE AND RELATED COMPLEXES

Yoshio Matsunaga

Department of Chemistry, Faculty of Science, Hokkaido

University, Sapporo, Japan

Phenoxazine (I), phenothiazine (II), and phenoselenazine (III) form solid iodine complexes of a 2:3 mole ratio when the components dissolved in benzene are mixed (1,2). They are all black poly-crystalline powder.

(I)   X = O

(II)   X = S

(III)   X = Se

The electrical resistivity of compressed samples shows a temperature dependence which follows the typical semiconductor behavior: $\rho = \rho_0 \exp(E/kT)$. The following values of the resistivity (ρ) at 20°C and the activation energy (E) in the higher temperature region have been obtained: the phenoxazine complex, 10 ohm cm, 0.14 eV; the phenothiazine complex, 19 ohm cm, 0.19 eV; the phenoselenazine complex, 32 ohm cm, 0.28 eV. The temperature dependence of the Seebeck coefficient (Q) is shown in Fig. 1 for these three complexes. At 20°C, the values are as follows: +5 μV/deg for the phenoxazine complex, +153 μV/deg for the phenothiazine complex, and +4 μV/deg for the phenoselenazine complex. The higher-temperature region of the curve is nearly straight in the cases of the phenoxazine and phenoselenazine complexes and seems to fit in with the equation of Johnson and Lark-Horowitz for the Seebeck coefficient of an intrinsic semiconductor. The ratio (b) of the electron mobility (μe) to the hole mobility is evaluated as 2.11 for the phenoxazine complex and as 1.07 for the phenoselenazine complex. Assuming that the density of states can be approximated by the number of organic molecules per unit volume, 2.54 x 10^{21} per cm^3

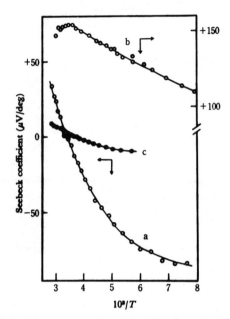

Fig. 1   Seebeck coefficients of the iodine complexes of a)
phenoxazine, b) phenothiazine, and c) phenoselenazine.

for the phenoxazine complex and 2.50×10^{21} per cm^3 for the phenoselenazine complex, the electron mobility is estimated to be 0.04
and 2.6 cm^2/V sec respectively.

The 2:3 iodine complexes of N-alkyl- and benzo-phenothiazines
have been found to behave similarly (2,3). The results are summarized in Table 1. The Seebeck coefficients of the benzophenothi-
azine complexes are nearly parallel to that for the phenoselenazine
complex; however, the linear parts are too short to fit the equation
of Johnson and Lark-Horowitz.

Table 1   Electrical properties of the iodine complexes of N-alkyl-
and benzo-phenothiazines.

Organic component	ρ at 20°C (ohm cm)	E (eV)	Q at 20°C (μV/deg)	b	μ_e (cm^2/V sec)
N-Methylphenothiazine	2.0	0.20	+10	2.08	2.3
N-Ethylphenothiazine	74	0.15	+57	0.20	10^{-3}
Benzo[a]phenothiazine	34	0.16	-47	>1	---
Benzo[c]phenothiazine	18	0.14	-57	>1	---

The vibrational spectrum provides a good method for the char-
acterization of the electronic structure of the organic molecules
in these iodine complexes. For example, the spectrum of the phenothiazine complex is compared with the spectrum of the parent organic

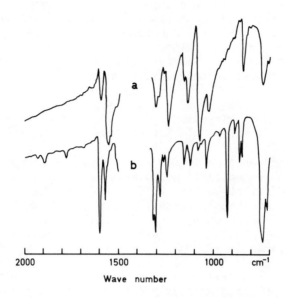

Fig. 2   Vibrational spectra of the phenothiazine–iodine complex (a) and phenothiazine itself (b).

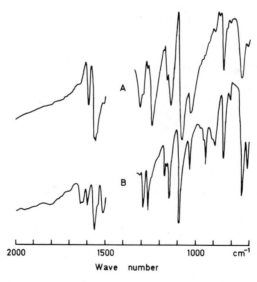

Fig. 3   Two extremes in the vibrational spectrum of semiquinoid phenothiazine bromide. The spectrum is often a superposition of these two.

compound in Fig. 2. It is easy to see that these two are distinctly
different from each other, but the former spectrum is very close to
the spectrum A of the semiquinoid bromide, a cation-radical salt
derived from phenothiazine, given in Fig. 3 (1,4). Therefore, it
may be concluded that every phenothiazine molecule in the iodine
complex bears an approximately unit-positive charge.

As the temperature dependence of Seebeck coefficient of the
phenothiazine complex given in Fig. 1 is not of an intrinsic semiconductor, we examined the electrical properties of the semiquinoid
iodide-iodine system (5). The monoiodide hitherto unknown was prepared by the reaction of an equimolar mixture of phenothiazine and
its S-oxide with concentrated hydroiodic acid (2). The resistivity
and Seebeck coefficient both at 20°C are plotted against the mole
ratio (total iodine/phenothiazine) in Fig. 4. A small resistivity
maximum is located at a mole ratio of 1.25, the value at this composition being about 800 ohm cm. Then the resistivity reaches a minimum at a mole ratio of 1.50 which is the composition of the phenothiazine complex deposited from solution. The Seebeck coefficient
changes its sign from negative to positive at a mole ratio of 1.25.
Furthermore, the activation energy for semiconduction decreases by

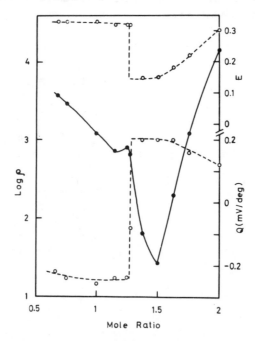

Fig. 4   Electrical resistivity (ρ), activation energy for semiconduction (E), and Seebeck coefficient (Q) at 20°C plotted against
the mole ratio (total iodine/phenothiazine) in the phenothiazine
iodide-iodine system.

a factor of two at this composition. These pieces of evidence strongly suggest that the phenothiazine-iodine system is an intrinsic semiconductor at a mole ratio of 1.25. The electron mobility estimated is about 0.9 cm²/V sec. The additional iodine must be divided into three fractions. First of all, 0.75 mole is to form an intrinsic semiconductor. Then, 0.25 mole of iodine makes the complex a semiconductor of the p-type. Finally, the iodine exceeding the mole ratio of 2:3 seems to play no essential role in the conduction mechanism. The partial molal volume of iodine is constant throughout the examined range.

In Fig. 5 the electronic spectra, as measured by the diffuse reflectance method, are shown for the iodine complexes of phenoxazine, phenothiazine, and phenoselenazine. The maximum of the low-energy band in the thiazine complex is found at 5.2 kK, while those of the other two complexes are expected to be below 4 kK, the limit of the spectroreflectometer employed. The appearance of a low-energy absorption band attributable to an electronic transition is common to low-resistivity organic semiconductors, e.g., the TCNQ anion-radical salts and some ionic molecular complexes. However, the location of its maximum is not necessarily correlated with the resistivity, as have been pointed out in our work on the anion-radical salts derived from tetrahalo-p-diphenoquinones (6). In the present complexes, it may be noted that the location of the maximum cannot be correlated with the magnitude of the activation energy.

We also examined the semiquinoid phenothiazine bromide-iodine system (4). A deep minimum of the resistivity is located around mole ratios (iodine/bromide) of 0.25 to 0.30, see Fig. 6. The adduct may be a bromide doped with iodine or a complex cation-radical salt, (phenothiazine)$_2$Br$_2$I, which is a cationic analogue of the TCNQ

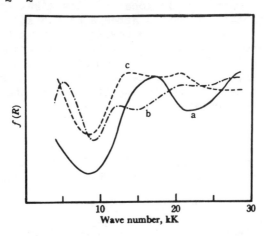

Fig. 5   Diffuse reflection spectra of the iodine complexes of a) phenoxazine, b) phenothiazine, and c) phenoselenazine.

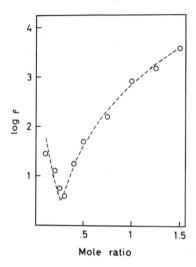

Fig. 6 Electrical resistivity plotted against the mole ratio (iodine/bromide) in the phenothiazine bromide-iodine system.

complex anion-radical salts well known for their low electrical resistivities, as Br_2I^- is a well-known species. Drastic changes in the vibrational and electronic spectra by the iodine content have been noted. The low-resistivity adduct shows a vibrational spectrum consisting of very broad lines. Nevertheless, the whole pattern is rather close to the spectrum A of the bromide in Fig. 3. The electronic spectrum bears also some resemblance to that of the bromide, especially the one showing the vibrational spectrum A. The maximum of the low-energy absorption is located below 4 kK. At a mole ratio of 0.40, an additional absorption appears at 5.5 kK. By further addition of iodine, the absorption having a maximum below 4 kK is completely replaced by the one at 5.5 kK. Then, around a mole ratio of 1.00, the intensity of the absorption at 5.5 kK is markedly diminished. Moreover, the vibrational spectrum is characterized by a superposition of a pattern consisting of lines as sharp as those in the spectrum B of the bromide.

(1) Y. Matsunaga, Helv. Phys. Acta, 36, 800 (1963).
(2) Y. Matsunaga and Y. Suzuki, Bull. Chem. Soc. Japan, 45, 3375 (1972); ibid., 46, 719 (1973).
(3) K. Kan and Y. Matsunaga, ibid., 45, 2096 (1972).
(4) Y. Matsunaga and K. Shono, ibid., 43, 2007 (1970).
(5) S. Doi, B.Sc. thesis, Hokkaido University (1973).
(6) Y. Matsunaga and Y. Narita, Bull. Chem. Soc. Japan, 45, 408 (1972).

AUTHOR INDEX

SUBJECT INDEX

1. Dr. S.Z. Weisz	17. Dr. H. Inokuchi	33. Dr. H. Kokado
2. Dr. C. Braun	18. Dr. A. Suna	34. Dr. M. Yokoyama
3. Dr. H. Mikawa	19. Dr. T.C. McGill	35. Dr. M. Kawabe
4. Dr. J. Mort	20. Dr. K. Iguchi	36. Miss E. Sakanaka
5. Dr. H. Akamatsu	21. Dr. J. Tanaka	37. Dr. P. Fielding
6. Dr. M. Silver	22. Dr. H. Kuroda	38. Dr. T. Kawakubo
7. Dr. K. Masuda	23. Dr. K. Taketani	39. Dr. Y. Takahashi
8. Dr. M. Pope	24. Dr. M. Mabuchi	40. Dr. K. Yoshino
9. Dr. S.I. Choi	25. Dr. N. Itoh	41. Dr. Y. Oeda
10. Dr. S. Takeno	26. Dr. Y. Harada	42. Dr. K. Okamoto
11. Dr. D.C. Hoesterey	27. Dr. A. Maruyama	43. Dr. K. Kato
12. Dr. W.A. Little	28. Dr. W.D. Gill	44. Dr. Y. Aoyagi
13. Dr. A. Bloch	29. Dr. K. Yoshihara	45. Mr. T. Ugumori
14. Dr. S.A. Rice	30. Dr. H. Yasunaga	46. Mrs. M. Sakurai
15. Dr. M. Sano	31. Dr. S. Kusabayashi	47. Mr. K. Hiroyama
16. Dr. U. Itoh	32. Dr. Y. Matsunaga	48. Mr. T. Nishimura